# Advanced Textbooks in Mathematics

ISSN: 2059-769X

*Published*

The Wigner Transform
  *by Maurice de Gosson*

*Forthcoming*

Periods and Special Functions in Transcendence
  *by Paula B Tretkoff*

Conformal Maps and Geometry
  *by Dmitry Belyaev*

Advanced Textbooks in Mathematics

# The Wigner Transform

## Maurice de Gosson

*University of Vienna, Austria*

**World Scientific**

NEW JERSEY · LONDON · SINGAPORE · BEIJING · SHANGHAI · HONG KONG · TAIPEI · CHENNAI · TOKYO

*Published by*

World Scientific Publishing Europe Ltd.

57 Shelton Street, Covent Garden, London WC2H 9HE

*Head office:* 5 Toh Tuck Link, Singapore 596224

*USA office:* 27 Warren Street, Suite 401-402, Hackensack, NJ 07601

**Library of Congress Cataloging-in-Publication Data**

Names: Gosson, Maurice de.
Title: The Wigner transform / by Maurice de Gosson (University of Vienna, Austria).
Description: New Jersey : World Scientific, 2017. | Series: Advanced textbooks in mathematics
Identifiers: LCCN 2016056680| ISBN 9781786343086 (hc : alk. paper) |
      ISBN 9781786343093 (pbk : alk. paper)
Subjects: LCSH: Phase space (Statistical physics) | Statistical mechanics.
Classification: LCC QC174.85.P48 G67 2017 | DDC 530.15/95--dc23
LC record available at https://lccn.loc.gov/2016056680

**British Library Cataloguing-in-Publication Data**

A catalogue record for this book is available from the British Library.

Desk Editors: V. Vishnu Mohan/Mary Simpson

Typeset by Stallion Press
Email: enquiries@stallionpress.com

Printed in Singapore

To My Charlyne, With All My Love

# Preface

The study of the Wigner transform and of its twin brother, the ambiguity function, has an already long and illustrious history, with strong links to both pure and applied mathematics, and to physics and engineering. In this book, we aim at giving a comprehensive and mathematically rigorous treatment of these subjects at a level which should be accessible to undergraduate upper-level students in mathematics and mathematical physics (we have put a special emphasize on the applications to quantum mechanics, some of them of a rather advanced level). The material should also be of interest to mathematicians and engineers working in signal analysis and time-frequency analysis; the confirmed researcher can use this book as a reference work, since we have tried to unify different topics which are more or less spread out in the literature.

Needless to say, this book is not the first approaching the Wigner transform using methods from harmonic analysis. Well-cited predecessors are the comprehensive books by Folland *Harmonic Analysis in Phase Space* and by Gröchenig *Time-Frequency Analysis*. The number of contributions in the form of research or expository papers defies the imagination: we have not been able to take into account all the recent advances, but rather focused on the potentially most important; of course these choices *do* reflect the author's own tastes.

The book is structured as a series of Lecture Notes; the order in which the topics are introduced is therefore linear. I have tried to present the material using the principle of parsimony, and the mathematics is presented

and structured following the old-fashioned but efficient way Definition —
Lemma — Proposition — Corollary (including, of course, Examples and
Remarks). The chapters are short, and designed for a 60 minutes lecture
(this may be wishful thinking — it depends of course on the initial level
of the audience; I have here in mind upper-level European or American
undergraduates or graduate students).

This work has been financed by the grant P27773-N23 of FWF Der
Wissenschaftsfonds (the Austrian Research Agency)

Maurice A. de Gosson
Vienna, Summer 2016

# Introduction

The Wigner transform has a long story which started in 1932 with Eugene Wigner's ground-breaking paper *On the quantum correction for thermodynamic equilibrium* (Phys. Rev. **40** (1932), 799–755). In this paper, Wigner introduced a probability quasi-distribution that allowed him to express quantum mechanical expectation values in the same form as the averages of classical statistical mechanics. There have been many speculations about how Wigner got his idea; his eponymous transform seems to be pulled out of thin air. In a footnote on the second page of his paper, Wigner writes *"This expression was found by L. Szilard and the present author some years ago for another purpose"*. However, the participation of Leo Szilard to this discovery is somewhat questionable; it is more likely that Wigner wanted to boost Szilard's career as the latter was leaving pre-Nazi Germany at this time. Truly, Wigner's definition, in modern notation

$$W\psi(x, p) = \frac{1}{2\pi\hbar} \int_{-\infty}^{\infty} e^{-\frac{i}{\hbar}py} \psi\left(x + \frac{1}{2}y\right) \overline{\psi\left(x - \frac{1}{2}y\right)} dy \qquad (0.1)$$

did not remind of anything one had seen before; a rapid glance suggests it is something of a mixture of a Fourier transform and of a convolution. And, yet, it worked! For instance, under some mild assumptions on the function $\psi$ we recover the probability amplitudes $|\psi(x)|^2$ and $|\phi(p)|^2$ of quantum mechanics:

$$\int_{-\infty}^{\infty} W\psi(x,p)dp = |\psi(x)|^2,$$

$$\int_{-\infty}^{\infty} W\psi(x,p)dx = |\phi(p)|^2;$$

assuming $\psi$ normalized to unity and integrating any of these equalities, we get

$$\int_{-\infty}^{\infty}\int_{-\infty}^{\infty} W\psi(x,p)dpdx = 1$$

so that the Wigner transform $W\psi$ can be used as a mock probability distribution. The rub comes from the fact that $W\psi(x,p)$ invariably takes negative values (except when $\psi$ is a Gaussian); it is therefore customary to call $W\psi$ a *quasi-distribution*. In the paper by Hillery *et al. Distribution Functions in Physics: Fundamentals* (Phys. Reps. 106 (1984), 121–167) — which is coauthored by Wigner himself — we are told that the particular choice (0.1) was made because it seemed to be the simplest of those for which each Galilei transformation corresponds to the same Galilei transformation of the quantum mechanical wave functions. In later work Wigner [1979] returned to this issue by considering properties which one would want such a distribution to satisfy. He then showed that the distribution given by (0.1) was the only one which satisfied these properties. A subsequent paper by O'Connell and Wigner *Some properties of a non-negative quantum-mechanical distribution function* (Phys. Lett. A 85(3), 121–126 (1981)) considered a somewhat different list of properties and showed that these, too, led to the expression in (0.1). The Wigner transform was rediscovered by Ville in signal analysis in 1948 and eventually became a popular tool among engineers through the work of Claasen and Mecklenbräuker; it is closely related to the windowed Fourier and Gabor transforms familiar in time-frequency analysis.

This book is organized as follows.

**Chapter 1:** We introduce two simple but essential unitary operators: the Heisenberg shift operator and the Grossmann–Royer parity operator. They will allow us not only to define the Wigner transform and the ambiguity function in a simple and natural way, but they will allow us to give neat and concise proofs of numerous important properties of the Wigner transform.

**Chapter 2:** We define the cross-Wigner transform using the Grossmann–Royer operator, and show that this definition is equivalent to

Wigner's original definition. We list and prove several important properties, and show that the Wigner transform can be extended to tempered distributions.

**Chapter 3:** The cross-ambiguity function is defined in terms of the Heisenberg operator; the usual explicit expression is derived. We prove that the cross-ambiguity function is closely related to the cross-Wigner transform, both analytically by a symplectic Fourier transform, and algebraically. The main properties are proven.

**Chapter 4:** In this chapter, we show that the Weyl operators familiar from the theory of pseudodifferential operators can be naturally defined using the cross-Wigner transform (or the cross-ambiguity function). We take the opportunity to discuss the various other possible definitions of Weyl operators. This chapter thus justifies the terminology "Weyl–Wigner–Moyal theory" which is often used.

**Chapter 5:** It is important to list the symmetries underlying a mathematical theory since they allow a better understanding of that theory and can considerably simplify its study. We show in this chapter that the natural group of symmetries in the Weyl–Wigner–Moyal theory is the symplectic group (this is the well-known symplectic covariance of the Wigner transform and the associated Weyl calculus). We show in addition that the symplectic group is a maximal group of symmetries.

**Chapter 6:** This chapter is devoted to an extremely useful formula, the Moyal identity (for both the cross-Wigner transform and the cross-ambiguity function). Not only does the Moyal identity to prove important regularity results, it is also an essential tool for proving inversion formula which allow the reconstruction of functions from the knowledge of their cross-Wigner transform (or their cross-ambiguity function). This has important applications to the study of the so-called weak values from time-symmetric quantum mechanics. The Moyal identity is also a key tool for the study of wavepacket transforms, generalizing the Bargmann transform.

**Chapter 7:** An essential property of the Wigner transform is that it allows the definition and the study of a fundamental functional space, the Feichtinger algebra, which is the simplest representative of the class of modulation spaces introduced by H. Feichtinger in the mid-1980's. In this chapter, we study and define this algebra, exploiting the properties of the Wigner transform in place of those of Gabor transform which is more usual in the literature. This approach allows to highlight in an easy way the metaplectic invariance of the Feichtinger algebra.

**Chapter 8:** The Wigner transform was introduced as a substitute for a probability density in phase space. It is however not the only possible choice. In this chapter we discuss the so-called *Cohen class* whose elements are obtained by convolution of the (cross-)Wigner distribution with a tempered distribution. We study two examples: the Husimi distribution and its generalization, and the Born–Jordan distribution, which is a newcomer.

**Chapter 9:** Gaussian functions play a privileged role in many parts of analysis. In this chapter we give explicit expressions for the cross-Wigner transform of generalized Gaussians (the " squeezed coherent states of quantum optics and mechanics"), as a particular result we recover the well-known formula for the Wigner transform of the standard Gaussian. We complete our results by computing explicitly the cross-Wigner transform of the Hermite functions.

**Chapter 10:** We study in this chapter the properties of functions whose Wigner transform is dominated by a phase space Gaussian function. We show how this problem is related to Hardy's uncertainty principle, which gives conditions for a function and its Fourier transform to have exponential decrease. This requires tools from symplectic geometry (Williamson's symplectic diagonalization theorem, and its variants).

**Chapter 11:** In this chapter (which is the first directly concerned with quantum-mechanical applications), we define in a rigorous way the notions of Moyal star-product, and of the associated twisted convolution. The Moyal star-product is an essential ingredient in deformation quantization. We show that the latter can be viewed as a phase space pseudodifferential calculus, intertwined with the standard Weyl calculus using the wavepacket transforms previously introduced.

**Chapter 12:** This chapter is devoted to the statistical aspects of quantum mechanics from the phase space point of view, and hence very much along the lines of Wigner. A novelty is however the definition of a study of the notion of quantum-mechanical weak values, which play a fundamental role in modern quantum physics (the time-symmetric theory, as it is also called).

**Chapter 13:** We define a study of the essential notion of density operator, which is closely associated to mixed states in quantum mechanics. The Weyl symbol of a density operator is a convex sum of (possibly infinitely many) Wigner transforms. Density operators are self-adjoint positive trace-class operators; we study these in some detail, having in mind the difficulties due to positivity issues. We prove the strong Robertson–Schrödinger uncertainty principle.

**Chapter 14:** We develop the theory of density operators initiated in Chapter 13 by focusing on the positivity issues, which are related to the verification of the Kastler–Loupias–Miracle Sole condition, which are a quantum version of Bochner's theorem on functions of positive type. This leads us to introduce the Narcowich–Wigner spectrum of a symbol, and to shortly discuss the open yet unsolved problems.

**Chapter 15:** In this last chapter we introduce, following previous work of ours, the geometric and topological notion of "quantum blob". A quantum blob is the image of a phase space ball with radius $\sqrt{\hbar}$, As simple as this definition may seem, it is closely related to the Wigner transform of Gaussians, and can be used to obtain a symplectically invariant coarse-graining of classical phase space. The notion is also closely related to the topological notion of symplectic capacity, which is itself related to Gromov's non-squeezing theorem.

# Contents

*Preface*                                                                    vii

*Introduction*                                                                ix

## Part I  General Mathematical Framework                                     1

1   Phase Space Translations and Reflections                                   3

   1.1   Some Notation . . . . . . . . . . . . . . . . . . . . . . . . . .    3
         1.1.1   The spaces $\mathbb{R}_x^n$ and $\mathbb{R}_p^n$ . . . . . . . . . . . . . . .   3
         1.1.2   The symplectic structure of phase space . . . . . .          4
         1.1.3   Some usual function spaces . . . . . . . . . . . . .          5
         1.1.4   Fourier transform . . . . . . . . . . . . . . . . . . .       6
   1.2   The Heisenberg–Weyl and Grossmann–Royer
         Operators . . . . . . . . . . . . . . . . . . . . . . . . . . . .     6
         1.2.1   The displacement Hamiltonian . . . . . . . . . . .            6
         1.2.2   The Heisenberg–Weyl operators . . . . . . . . . .             7
         1.2.3   The Grossmann–Royer parity operators . . . . . .             9
   1.3   A Functional Relation Between $\widehat{T}(z_0)$ and $\widehat{R}(z_0)$ . . . . .  12
   1.4   Quantization of Exponentials . . . . . . . . . . . . . . . .         14

xv

2    The Cross-Wigner Transform                                      17

     2.1    Definitions of the Cross-Wigner Transform . . . . . . . . .   17
            2.1.1   First definition . . . . . . . . . . . . . . . . . .   17
            2.1.2   Wigner's definition . . . . . . . . . . . . . . . . .  18
            2.1.3   The Gabor transform and its variants . . . . . . .    19
            2.1.4   Extension to tempered distributions . . . . . . . .   20
     2.2    Properties of the Cross-Wigner Transform . . . . . . . . .    22
            2.2.1   Elementary algebraic properties  . . . . . . . . . .  22
            2.2.2   Analytical properties and continuity . . . . . . . .  24
            2.2.3   The marginal properties  . . . . . . . . . . . . . .  26
            2.2.4   Translating Wigner transforms . . . . . . . . . . .   27

3    The Cross-Ambiguity Function                                    31

     3.1    Definition of the Cross-Ambiguity Function . . . . . . . .    31
            3.1.1   Definition using the Heisenberg–Weyl
                    operator  . . . . . . . . . . . . . . . . . . . . . . 31
            3.1.2   Traditional definition  . . . . . . . . . . . . . .   32
            3.1.3   The Fourier–Wigner transform . . . . . . . . . . .    33
     3.2    Properties and Relation with the Wigner Transform  . . .     34
            3.2.1   Properties of the cross-ambiguity function . . . . .  34
            3.2.2   Relation with the cross-Wigner transform . . . . .    35
            3.2.3   The maximum of the ambiguity function  . . . . .      37

4    Weyl Operators                                                  39

     4.1    The Notion of Weyl Operator . . . . . . . . . . . . . . . .   39
            4.1.1   Weyl's definition, and rigorous definitions . . . . . 39
            4.1.2   The distributional kernel of a Weyl operator . . .    43
            4.1.3   Relation with the cross-Wigner transform . . . . .    47
     4.2    Some Properties of the Weyl Correspondence . . . . . . .     49
            4.2.1   The adjoint of a Weyl operator . . . . . . . . . . .  49
            4.2.2   An $L^2$ boundedness result . . . . . . . . . . . . . 50

5    Symplectic Covariance                                           53

     5.1    Symplectic Covariance Properties . . . . . . . . . . . . . .  53
            5.1.1   Review of some properties of $\mathrm{Mp}(n)$
                    and $\mathrm{Sp}(n)$ . . . . . . . . . . . . . . . . .53
            5.1.2   Proof of the symplectic covariance property . . . .  55

5.1.3 Symplectic covariance of Weyl operators . . . . . 58
5.2 Maximal Covariance . . . . . . . . . . . . . . . . . . . . 59
5.2.1 Antisymplectic matrices . . . . . . . . . . . . . . 60
5.2.2 The maximality property . . . . . . . . . . . . . 62
5.2.3 The case of Weyl operators . . . . . . . . . . . . 64

6 The Moyal Identity                                          67

6.1 Precise Statement and Proof . . . . . . . . . . . . . . . 67
6.1.1 The general Moyal identity . . . . . . . . . . . . 67
6.1.2 A continuity result . . . . . . . . . . . . . . . . . 69
6.2 Reconstruction Formulas . . . . . . . . . . . . . . . . . 70
6.2.1 Reconstruction using the cross-Wigner
transform . . . . . . . . . . . . . . . . . . . . . . 70
6.2.2 Reconstruction using the cross-ambiguity
function . . . . . . . . . . . . . . . . . . . . . . . 71
6.3 The Wavepacket Transforms . . . . . . . . . . . . . . . 72
6.3.1 Definition . . . . . . . . . . . . . . . . . . . . . . 72
6.3.2 Properties of the wavepacket transform . . . . . . 73

7 The Feichtinger Algebra                                     77

7.1 Definition and First Properties . . . . . . . . . . . . . . 77
7.1.1 Definition of $S_0(\mathbb{R}^n)$ . . . . . . . . . . . . . . . . 77
7.1.2 Analytical properties of $S_0(\mathbb{R}^n)$ . . . . . . . . . 82
7.1.3 The algebra property of $S_0(\mathbb{R}^n)$ . . . . . . . . . 85
7.2 The Dual Space $S_0'(\mathbb{R}^n)$ . . . . . . . . . . . . . . . . . 86
7.2.1 Description of $S_0'(\mathbb{R}^n)$ . . . . . . . . . . . . . . . 86
7.2.2 The Gelfand triple $(S_0, L^2, S_0')$ . . . . . . . . . . 87

8 The Cohen Class                                             89

8.1 Definition . . . . . . . . . . . . . . . . . . . . . . . . . . 89
8.1.1 The marginal conditions . . . . . . . . . . . . . . 92
8.1.2 Generalization of Moyal's identity . . . . . . . . 94
8.1.3 The operator calculus associated with $Q$ . . . . . 95
8.2 Two Examples . . . . . . . . . . . . . . . . . . . . . . . . 97
8.2.1 The generalized Husimi distribution . . . . . . . . 97
8.2.2 The Born–Jordan transform . . . . . . . . . . . 100

9   Gaussians and Hermite Functions                                    103

    9.1   Wigner Transform of Generalized Gaussians   . . . . . . .   103
          9.1.1   Generalized Gaussian functions . . . . . . . . . . .   103
          9.1.2   Explicit results . . . . . . . . . . . . . . . . . . .   105
          9.1.3   Cross-ambiguity function of a Gaussian . . . . . .   108
          9.1.4   Hudson's theorem   . . . . . . . . . . . . . . . . .   109
    9.2   The Case of Hermite Functions   . . . . . . . . . . . . .   109
          9.2.1   Short review of the Hermite and Laguerre
                  functions   . . . . . . . . . . . . . . . . . . . . .   109
          9.2.2   The Wigner transform of Hermite functions . . . .   112
          9.2.3   The cross-Wigner transform of Hermite
                  functions   . . . . . . . . . . . . . . . . . . . . .   113
          9.2.4   Flandrin's conjecture   . . . . . . . . . . . . . . .   116

10  Sub-Gaussian Estimates                                             119

    10.1  Hardy's Uncertainty Principle . . . . . . . . . . . . . . .   119
          10.1.1  The one-dimensional case . . . . . . . . . . . . .   119
          10.1.2  Two lemmas   . . . . . . . . . . . . . . . . . . . .   120
          10.1.3  The multidimensional Hardy uncertainty
                  principle   . . . . . . . . . . . . . . . . . . . . .   121
    10.2  Sub-Gaussian Estimates for the Wigner Transform   . . . .   123
          10.2.1  Statement of the result   . . . . . . . . . . . . .   123
          10.2.2  First proof . . . . . . . . . . . . . . . . . . . . .   124
          10.2.3  Second proof . . . . . . . . . . . . . . . . . . . .   126

## Part II   Applications to Quantum Mechanics                         **129**

11  Moyal Star Product and Twisted Convolution                        131

    11.1  The Moyal Product of Two Symbols . . . . . . . . . . . .   131
          11.1.1  Definition of the Moyal product   . . . . . . . . .   131
          11.1.2  Twisted convolution . . . . . . . . . . . . . . . .   134
    11.2  Bopp Operators . . . . . . . . . . . . . . . . . . . . . .   136
          11.2.1  Bopp shifts . . . . . . . . . . . . . . . . . . . . .   136
          11.2.2  Definition and justification of Bopp
                  operators   . . . . . . . . . . . . . . . . . . . . .   139
          11.2.3  The intertwining property . . . . . . . . . . . . .   141

12 Probabilistic Interpretation of the Wigner Transform 143

12.1 Introduction . . . . . . . . . . . . . . . . . . . . . . . 143
12.1.1 Back to Wigner . . . . . . . . . . . . . . . . . 143
12.1.2 Averaging observables and symbols . . . . . . . 144
12.2 The Strong Uncertainty Principle . . . . . . . . . . . . 146
12.2.1 Variances and covariances . . . . . . . . . . . . 146
12.2.2 The uncertainty principle . . . . . . . . . . . . 147
12.2.3 The quantum covariance matrix . . . . . . . . . 148
12.3 The Notion of Weak Value . . . . . . . . . . . . . . . . 150
12.3.1 Definition of weak values . . . . . . . . . . . . 150
12.3.2 A complex phase space distribution . . . . . . . 152
12.3.3 Reconstruction using weak values . . . . . . . . 152

13 Mixed Quantum States and the Density Operator 155

13.1 Trace Class Operators . . . . . . . . . . . . . . . . . . 155
13.1.1 Definition and general properties . . . . . . . . 156
13.1.2 The case of Weyl operators . . . . . . . . . . . 157
13.2 The Density Operator . . . . . . . . . . . . . . . . . . 159
13.2.1 The Wigner transform of a mixed state . . . . . 159
13.2.2 A characterization of density operators . . . . . 161
13.2.3 Uncertainty principle for density operators . . . . 162
13.2.4 Covariance matrix . . . . . . . . . . . . . . . . 165

14 The KLM Conditions and the Narcowich–Wigner Spectrum 169

14.1 The Quantum Bochner Theorem . . . . . . . . . . . . . 169
14.1.1 Bochner's theorem . . . . . . . . . . . . . . . . 169
14.1.2 The quantum case: the KLM conditions . . . . . 171
14.1.3 The quantum covariance matrix . . . . . . . . . 175
14.2 The Narcowich–Wigner Spectrum . . . . . . . . . . . . 179
14.2.1 $\eta$-Positive functions . . . . . . . . . . . . . . . 179
14.2.2 The Narcowich–Wigner spectrum
of some states . . . . . . . . . . . . . . . . . 180

15 Wigner Transform and Quantum Blobs 183

15.1 Quantum Blobs and Phase Space . . . . . . . . . . . . 183
15.1.1 Geometric definition of a quantum blob . . . . . 184
15.1.2 Quantum phase space . . . . . . . . . . . . . . 185

15.2　Quantum Blobs and the Wigner Transform ........ 186

　　　15.2.1　The basic example .................. 186

　　　15.2.2　Covariance ellipsoid and quantum blobs ...... 187

15.3　From One Quantum Blob to Another ........... 189

　　　15.3.1　The general case .................. 190

　　　15.3.2　Averaging over quantum blobs ........... 192

**Appendix A　$Sp(n)$ and $Mp(n)$**　　　　　　　　　　　　195

　A.1　The Symplectic Group ..................... 195

　A.2　The Metaplectic Group .................... 198

　A.3　The Inhomogeneous Metaplectic Group ......... 200

**Appendix B　The Symplectic Fourier Transform**　　　203

**Appendix C　Symplectic Diagonalization**　　　　　　207

　C.1　Williamson's Theorem ................... 207

　C.2　The Block-Diagonal Case ................. 209

　C.3　The Symplectic Case .................... 210

　C.4　The Symplectic Spectrum ................. 211

**Appendix D　Symplectic Capacities**　　　　　　　　215

　D.1　Gromov's Non-squeezing Theorem ............ 215

　D.2　Symplectic Capacities ................... 216

　D.3　Properties ......................... 217

　D.4　The Symplectic Capacity of an Ellipsoid ......... 218

*Bibliography*　　　　　　　　　　　　　　　　　　221

*Index*　　　　　　　　　　　　　　　　　　　　227

# Part I
# General Mathematical Framework

Chapter 1

# Phase Space Translations and Reflections

**Summary 1.** The Heisenberg–Weyl and Grossmann–Royer operators are elementary unitary operators corresponding, respectively, to translations and to reflections about a phase space point; these operators can be viewed as exponentials of certain first-order differential operators. They are related to the notion of quantization of a symbol (= generalized classical observable), and will allow a simple definition of the Wigner and ambiguity transforms.

## 1.1. Some Notation

In this preliminary section, we introduce some basic notation as well as the definition of the usual function spaces we will use.

### 1.1.1. The spaces $\mathbb{R}_x^n$ and $\mathbb{R}_p^n$

We denote by $\mathbb{R}_x^n$ the space of all vectors $x = (x_1, \ldots, x_n)$ interpreted as "generalized positions" and by $\mathbb{R}_p^n$ the space of all vectors $p = (p_1, \ldots, p_n)$, viewed as "generalized momenta"; of course the distinction between both vectors spaces $\mathbb{R}_x^n$ and $\mathbb{R}_p^n$ is purely notational, and we will not need to give it any deeper meaning (in mathematics, one would often view $\mathbb{R}_p^n$ as the algebraic dual of $\mathbb{R}_x^n$, but this approach is not useful unless one works on curved configurations spaces). The vector space $\mathbb{R}_x^n$ is equipped with the usual scalar product $x \cdot x' = x_1 x_1' + \cdots + x_n \cdot x_n'$; the scalar product $p \cdot p'$ on $\mathbb{R}_p^n$ is defined in a similar way (we will occasionally drop the dots and write

3

simply $xx'$ and $pp'$). Most of the time we will not distinguish between $\mathbb{R}_x^n$ and $\mathbb{R}_p^n$ and write simply $\mathbb{R}^n$.

We will denote by $\mathbb{R}_z^{2n}$ or simply $\mathbb{R}^{2n}$ the corresponding phase space, viewed as the Cartesian product $\mathbb{R}_x^n \times \mathbb{R}_p^n$: it consists of all $2n$ vectors $z = (x, p)$.

### 1.1.2. The symplectic structure of phase space

The phase space $\mathbb{R}^{2n}$ is equipped with two natural structures: the scalar product $z \cdot z' = x \cdot x' + p \cdot p'$ and the standard symplectic structure $\sigma$, defined by the standard symplectic form

$$\sigma(z, z') = p \cdot x' - p' \cdot x.$$

We will most of the time identify $\mathbb{R}_x^n \times \mathbb{R}_p^n$ with $\mathbb{R}^n \times \mathbb{R}^n \equiv \mathbb{R}^{2n}$. *A remark:* we have been writing vectors $x, p, z$ as line vectors because they are simpler to do so typographically; in matrix calculations, these vectors should be viewed as column vectors to avoid inconsistencies. For instance, introducing the "standard symplectic matrix"

$$J = \begin{pmatrix} 0 & I \\ -I & 0 \end{pmatrix}$$

($0$ and $I$ the $n \times n$ zero and identity matrices), the symplectic form $\sigma$ is given by

$$\sigma(z, z') = Jz \cdot z',$$

which means

$$\sigma(z, z') = z'^T J z = \begin{pmatrix} x' & p' \end{pmatrix} J \begin{pmatrix} x \\ p \end{pmatrix}$$

in matrix notation.

The symplectic group $\mathrm{Sp}(n)$ is the group of all linear automorphisms of $\mathbb{R}^{2n}$ preserving the symplectic form: we have $S \in \mathrm{Sp}(n)$ if and only if $\sigma(Sz, Sz') = \sigma(z, z')$ for all $(z, z') \in \mathbb{R}^{2n} \times \mathbb{R}^{2n}$.

We list here the essentials of the theory of the symplectic group; a more thorough treatment is given in Appendix A. A $2n \times 2n$ real matrix $S$ is called a symplectic matrix if we have $\sigma(Sz, Sz') = \sigma(z, z')$ for all vectors $z, z'$ in the phase space $\mathbb{R}_z^{2n}$. It is easy to check that $S$ is symplectic if it satisfies any of the two equivalent conditions

$$S^T J S = J, \quad S J S^T = J,$$

where $S^T$ is the transpose of $S$. Symplectic matrices form a group: first, if $S$ and $S'$ are symplectic, then

$$(SS')^T J(SS') = S'^T (S^T JS)S' = S'^T JS' = J,$$

hence $SS'$ is symplectic as well. Secondly, we have

$$(S^T JS)^{-1} = S^{-1} J^{-1} (S^T)^{-1} = J^{-1};$$

since $J^{-1} = -J$, we thus have $S^{-1} J(S^T)^{-1} = J$, hence $S^{-1}$ is symplectic. The group of all $2n \times 2n$ symplectic matrices is called the symplectic group, and it is denoted by $\mathrm{Sp}(n)$ (one also commonly finds the notation $\mathrm{Sp}(2n)$ and $\mathrm{Sp}(2n, \mathbb{R})$ in the literature).

### 1.1.3. Some usual function spaces

We denote by $\mathcal{S}(\mathbb{R}^m)$ the complex vector space of Schwartz test functions: $\psi \in \mathcal{S}(\mathbb{R}^m)$ if $\psi$ is infinitely differentiable, and if the functions $x^\beta \partial_x^\alpha \psi$ are bounded for all multi-indices $\alpha = (\alpha_1, \ldots, \alpha_m)$ and $\beta = (\beta_1, \ldots, \beta_m)$ in $\mathbb{N}^m$; by definition, $x^\beta = x_1^{\beta_1} \cdots x_m^{\beta_m}$ and $\partial_x^\alpha = \partial_{x_1}^{\alpha_1} \cdots \partial_{x_m}^{\alpha_m}$. The space $\mathcal{S}(\mathbb{R}^m)$ is a Fréchet space for the seminorms

$$\|\psi\|_{\alpha,\beta} = \sup |x^\beta \partial_x^\alpha \psi|; \tag{1.1}$$

using the generalized Leibniz formula, one shows that an equivalent system of seminorms on $\mathcal{S}(\mathbb{R}^m)$ is given by

$$\|\psi\|'_{\alpha,\beta} = \sup |\partial_x^\alpha x^\beta \psi|. \tag{1.2}$$

The dual space of $\mathcal{S}(\mathbb{R}^m)$ is denoted by $\mathcal{S}'(\mathbb{R}^m)$; its elements are called tempered distributions on $\mathbb{R}^m$. The distributional pairing between $\psi \in \mathcal{S}'(\mathbb{R}^m)$ and $\phi \in \mathcal{S}(\mathbb{R}^m)$ is denoted by $\langle \psi, \phi \rangle$. When $\psi$ and $\phi$ are both in $\mathcal{S}(\mathbb{R}^m)$ then

$$\langle \psi, \phi \rangle = \int_{\mathbb{R}^n} \psi(x)\phi(x)dx.$$

We denote by $L^2(\mathbb{R}^m)$ the Hilbert space of all (classes of) square-integrable complex functions on $\mathbb{R}^m$; it is equipped with the scalar product

$$(f|g)_{L^2(\mathbb{R}^m)} = \int_{\mathbb{R}^m} f(x)\overline{g(x)}dx.$$

When no confusion is likely to arise, we write simply $(f|g)_{L^2}$. The scalar product $(f|g)_{L^2}$ is related to the physicist's "bra-ket" product $\langle f|g \rangle$ by conjugation:

$$\langle f|g \rangle = \overline{(f|g)_{L^2}}.$$

The natural inclusions

$$\mathcal{S}(\mathbb{R}^m) \subset L^2(\mathbb{R}^m) \subset \mathcal{S}'(\mathbb{R}^m)$$

are continuous and $\mathcal{S}(\mathbb{R}^m)$ is dense in $L^2(\mathbb{R}^m)$.

### 1.1.4. Fourier transform

The Fourier transform on $\mathbb{R}^m$ is denoted by $F$; we will use the $\hbar$-dependent definition

$$Ff(x) = \left(\frac{1}{2\pi\hbar}\right)^{m/2} \int_{\mathbb{R}^n} e^{-\frac{i}{\hbar}x\cdot x'} f(x')dx'.$$

It is an automorphism $\mathcal{S}(\mathbb{R}^m) \longrightarrow \mathcal{S}(\mathbb{R}^m)$ which extends into a unitary automorphism of $L^2(\mathbb{R}^m)$ and whose inverse is given by

$$F^{-1}f(x) = \left(\frac{1}{2\pi\hbar}\right)^{m/2} \int_{\mathbb{R}^n} e^{\frac{i}{\hbar}x\cdot x'} f(x')dx'.$$

It also extends by duality into an automorphism of $\mathcal{S}'(\mathbb{R}^{nm})$.

The Fourier transform satisfies the Parseval formula

$$\int_{\mathbb{R}^m} Ff(x)\overline{Fg(x)}dx = \int_{\mathbb{R}^m} f(x)\overline{g(x)}dx$$

for $f, g \in L^2(\mathbb{R}^m)$.

## 1.2. The Heisenberg–Weyl and Grossmann–Royer Operators

### 1.2.1. The displacement Hamiltonian

For each $z_0 = (x_0 p_0)$ we define a Hamiltonian function $H_{z_0}$ by

$$H_{z_0}(z) = \sigma(z, z_0) = p \cdot x_0 - p_0 \cdot x.$$

The solutions of the associated Hamilton equations $\dot{x} = x_0$, $\dot{p} = p_0$ are

$$x(t) = x(0) + tx_0, \quad p(t) = p(0) + tp_0,$$

which we write

$$z(t) = T(tz_0)z(0),$$

where $T(z_0)$ is the phase space translation operator in the direction $z_0$. Let us look at the Schrödinger equation corresponding to the translation Hamiltonian; it is

$$i\hbar \frac{\partial \psi}{\partial t} = \widehat{H}_{z_0}\psi, \quad \psi(x,0) = \psi_0(x), \tag{1.3}$$

where $\widehat{H}_{z_0}$ is the operator obtained from $H_{z_0}$ by formally replacing $p$ by $-i\hbar\partial_x$:

$$\widehat{H}_{z_0} = \sigma(\hat{z}, z_0) = -i\hbar x_0 \cdot \partial_x - p_0 \cdot x. \tag{1.4}$$

The solution of (1.3) can be formally written as

$$\psi(x,t) = \widehat{T}(z_0,t)\psi_0(x) = e^{-\frac{it}{\hbar}\sigma(\hat{z},z_0)}\psi_0(x). \tag{1.5}$$

Using for instance the method of characteristics, one checks that this solution is explicitly given by the formula

$$\widehat{T}(z_0,t)\psi_0(x) = e^{\frac{i}{\hbar}(tp_0 \cdot x - \frac{1}{2}t^2 p_0 \cdot x_0)}\psi_0(x - tx_0). \tag{1.6}$$

It is clear that $\widehat{T}(z_0,t)$ is a unitary operator on $L^2(\mathbb{R}^n)$: we have

$$\|\widehat{T}(z_0,t)\psi\|_{L^2} = \|\psi\|_{L^2} \tag{1.7}$$

for every $\psi \in L^2(\mathbb{R}^n)$.

### 1.2.2. The Heisenberg–Weyl operators

Consider now the time-one solutions of the Schrödinger equation (1.3).

**Definition 2.** The operator $\widehat{T}(z_0) = \widehat{T}(z_0,1)$ is called the Heisenberg–Weyl operator, or displacement operator. Its action on a function $\psi$ is given by

$$\widehat{T}(z_0)\psi(x) = e^{\frac{i}{\hbar}(p_0 \cdot x - \frac{1}{2}p_0 \cdot x_0)}\psi(x - x_0). \tag{1.8}$$

$\widehat{T}(z_0)$ is also sometimes called "Weyl operator". We avoid this terminology since it is maybe confusing since we will use the term "Weyl operator" in a more general setting; see Chapter 4.

The Heisenberg–Weyl operators satisfy the commutation relation

$$\widehat{T}(z_0)\widehat{T}(z_1) = e^{\frac{i}{\hbar}\sigma(z_0,z_1)}\widehat{T}(z_1)\widehat{T}(z_0) \tag{1.9}$$

and the addition relation

$$\widehat{T}(z_0 + z_1) = e^{-\frac{i}{2\hbar}\sigma(z_0,z_1)}\widehat{T}(z_0)\widehat{T}(z_1) \tag{1.10}$$

for all $z_0, z_1 \in \mathbb{R}^{2n}$.

The unitarity of the operator $\widehat{T}(z_0)$ on $L^2(\mathbb{R}^n)$ is obvious; we have

$$(\widehat{T}(z_0)\psi|\widehat{T}(z_0)\phi)_{L^2} = \int_{\mathbb{R}^n} \psi(x-x_0)\overline{\phi(x-x_0)}dx = (\psi|\phi)_{L^2}.$$

Notice that

$$\widehat{T}(z_0)^{-1} = \widehat{T}(z_0)^* = \widehat{T}(-z_0).$$

The Heisenberg–Weyl operator can be extended to $\mathcal{S}'(\mathbb{R}^n)$ by the formula

$$\langle \widehat{T}(x_0,p_0)\psi, \phi \rangle = \langle \psi, \widehat{T}(-x_0,p_0)\phi \rangle, \tag{1.11}$$

where $\phi \in \mathcal{S}(\mathbb{R}^n)$. This formula gives the desired extension that follows from the fact that if $\psi \in \mathcal{S}(\mathbb{R}^n)$ then

$$\begin{aligned}
\langle \widehat{T}(x_0,p_0)\psi, \phi \rangle &= \int_{\mathbb{R}^n} e^{\frac{i}{\hbar}(p_0 \cdot x - \frac{1}{2}p_0 \cdot x_0)}\psi(x-x_0)\phi(x)dx \\
&= \int_{\mathbb{R}^n} e^{\frac{i}{\hbar}(p_0 \cdot x + \frac{1}{2}p_0 \cdot x_0)}\psi(x)\phi(x+x_0)dx \\
&= \int_{\mathbb{R}^n} \psi(x)\widehat{T}(-x_0,p_0)\phi(x)dx \\
&= \langle \psi, \widehat{T}(-x_0,p_0)\phi \rangle.
\end{aligned}$$

Now we have an important continuity property.

**Proposition 3.** (i) *The mapping* $z_0 \longmapsto \widehat{T}(z_0)$ *is strongly continuous on* $\mathcal{S}(\mathbb{R}^n)$, *i.e.*

$$\lim_{z \to z_0} \|\widehat{T}(z)\psi - \widehat{T}(z_0)\psi\|'_{\alpha,\beta} = 0$$

*for every* $\psi \in \mathcal{S}(\mathbb{R}^n)$, *and* $\alpha, \beta \in \mathbb{N}^n$, *where* $\|\cdot\|'_{\alpha,\beta}$ *is the seminorm* (1.2) *on* $\mathcal{S}(\mathbb{R}^n)$.
(ii) *It is weakly $*$-continuous on* $\mathcal{S}'(\mathbb{R}^n)$, *i.e.*

$$\lim_{z \to z_0} \langle \widehat{T}(z)\psi, \phi \rangle = \langle \widehat{T}(z_0)\psi, \phi \rangle$$

*for all* $\psi \in \mathcal{S}'(\mathbb{R}^n)$ *and* $\phi \in \mathcal{S}(\mathbb{R}^n)$.

**Proof.** (i) We are following [39, §9.5.1] (also see [52]). It is sufficient to assume that $z_0 = 0$; hence we have to prove that

$$\lim_{|z| \to 0} \|\partial_x^\alpha x^\beta(\widehat{T}(z)\psi - \psi)\|_\infty = 0.$$

Noting that by Leibniz' formula for the derivatives of a product we can write

$$\partial_x^\alpha x^\beta \widehat{T}(z)\psi = \sum_{\gamma \le \alpha, \delta \le \beta} c_{\alpha\beta\gamma\delta} x^\delta p^\gamma \widehat{T}(z)(\partial_x^{\alpha-\gamma} x^{\beta-\delta} \psi), \tag{1.12}$$

where the $c_{\alpha\beta\gamma\delta}$ are complex constants and $\gamma \le \alpha$ means $\gamma_j \le \alpha_j$ for $j = 1, 2, \ldots, n$, and we have

$$\|\partial_x^\alpha x^\beta (\widehat{T}(z)\psi - \psi)\|_\infty \le \|(\widehat{T}(z)(\partial_x^\alpha x^\beta \psi) - (\partial_x^\alpha x^\beta \psi)\|_\infty$$
$$+ \sum_{\substack{0 < \gamma \le \alpha \\ 0 < \delta \le \beta}} c_{\alpha\beta\gamma\delta} |x^\delta p^\gamma| \|\widehat{T}(z)(\partial_x^{\alpha-\gamma} x^{\beta-\delta} \psi)\|_\infty.$$

Now, since $\gamma \ne 0$, $\delta \ne 0$, it is clear that

$$\lim_{z \to 0} |x^\delta p^\gamma| \|\widehat{T}(z)(\partial_x^{\alpha-\gamma} x^{\beta-\delta} \psi)\|_\infty = 0$$

so there remains to show that

$$\lim_{z \to 0} \|(\widehat{T}(z)(\partial_x^\alpha x^\beta \psi) - (\partial_x^\alpha x^\beta \psi)\|_\infty = 0.$$

This equality is clear if $\psi$ is compactly supported, i.e. $\psi \in C_0^\infty(\mathbb{R}^n)$. The equality in the general case then follows from the density of $C_0^\infty(\mathbb{R}^n)$ in $\mathcal{S}(\mathbb{R}^n)$. (ii) It suffices again to assume $z_0 = 0$. Let $\psi \in \mathcal{S}'(\mathbb{R}^n)$ and $\phi \in \mathcal{S}(\mathbb{R}^n)$. By formula (1.11) we have

$$\lim_{z \to 0} \langle \widehat{T}(x,p)\psi, \phi \rangle = \lim_{z \to 0} \langle \psi, \widehat{T}(-x,p)\phi \rangle = \langle \psi, \phi \rangle,$$

and hence $\widehat{T}(z)$ is weakly $*$-continuous on $\mathcal{S}'(\mathbb{R}^n)$ as claimed. $\qquad\square$

### 1.2.3. The Grossmann–Royer parity operators

The Grossmann–Royer operators are essentially phase space reflection operators, and can be defined in terms of the Heisenberg–Weyl operators.

**Definition 4.** The Grossmann–Royer parity (or reflection) operator $\widehat{R}(z_0)$ is the operator

$$\widehat{R}(z_0) : \mathcal{S}(\mathbb{R}^n) \longrightarrow \mathcal{S}(\mathbb{R}^n)$$

defined by the formulas

$$\widehat{R}(z_0) = \widehat{T}(z_0)\widehat{R}(0)\widehat{T}(z_0)^{-1} \tag{1.13}$$

and

$$\widehat{R}(0)\psi(x) = \psi(-x). \tag{1.14}$$

The following properties are straightforward consequences of the definition.

**Proposition 5.** (i) *The Grossmann–Royer operators are linear involutions of $S(\mathbb{R}^n)$ and of $S'(\mathbb{R}^n)$; they are unitary on $L^2(\mathbb{R}^n)$ and the action of $\widehat{R}(z_0)$ (with $z_0 = (x_0, p_0)$) is explicitly given by the formula*

$$\widehat{R}(z_0)\psi(x) = e^{\frac{2i}{\hbar}p_0 \cdot (x-x_0)}\psi(2x_0 - x) \qquad (1.15)$$

*for any function (or distribution) $\psi : \mathbb{R}^n \longrightarrow \mathbb{C}$.*

(ii) *The Grossmann–Royer operators satisfy the product formula*

$$\widehat{R}(z_0)\widehat{R}(z_1) = e^{-\frac{2i}{\hbar}\sigma(z_0, z_1)}\widehat{T}(2(z_0 - z_1)). \qquad (1.16)$$

**Proof.** (i) The linearity of $\widehat{R}(z_0)$ is obvious. That $\widehat{R}(z_0)$ is an involution, i.e.

$$\widehat{R}(z_0)\widehat{R}(z_0) = I_{\mathrm{d}}$$

follows from the sequence of equalities

$$
\begin{aligned}
\widehat{R}(z_0)\widehat{R}(z_0) &= \widehat{T}(z_0)\widehat{R}(0)\widehat{T}(z_0)^{-1}\widehat{T}(z_0)\widehat{R}(0)\widehat{T}(z_0)^{-1} \\
&= \widehat{T}(z_0)\widehat{R}(0)\widehat{R}(0)\widehat{T}(z_0)^{-1} \\
&= \widehat{T}(z_0)\widehat{T}(z_0)^{-1} \\
&= I_{\mathrm{d}}.
\end{aligned}
$$

Setting $x' = 2x_0 - x$, we have

$$\|\widehat{R}(z_0)\psi\|_{L^2}^2 = \int_{\mathbb{R}^n} |\psi(2x_0 - x)|^2 dx = \|\psi\|_{L^2}^2;$$

hence $\widehat{R}(z_0)$ is also unitary (this property also follows directly from the unitarity of the Heisenberg–Weyl operators and of the parity operator $\widehat{R}(0)$). Formula (1.15) follows from the definition (1.13) by a straightforward calculation. (ii) We have, by formula (1.15),

$$
\begin{aligned}
\widehat{R}(z_0)\widehat{R}(z_1)\psi(x) &= \widehat{R}(z_0)[e^{\frac{2i}{\hbar}p_1 \cdot (2x_0 - x - x_1)}\psi(2x_1 - x)] \\
&= e^{\frac{2i}{\hbar}p_0 \cdot (x-x_0)}e^{\frac{2i}{\hbar}p_1 \cdot (x-x_1)}\psi(2x_1 - (2x_0 - x)) \\
&= e^{\frac{i}{\hbar}\Phi}\psi(x - 2(x_0 - x_1)),
\end{aligned}
$$

where the phase $\Phi$ is given by

$$\Phi = 2[(p_0 - p_1)x - p_0 x_0 - p_1 x_1 + 2p_1 x_0].$$

We also have by (1.15)

$$\widehat{T}(2(z_0 - z_1))\psi(x) = e^{\frac{i}{\hbar}\Phi'}\psi(x - 2(x_0 - x_1)),$$

where $\Phi'$ is given by

$$\Phi' = 2((p_0 - p_1) \cdot x - (p_0 - p_1) \cdot (x_0 - x_1)).$$

A straightforward calculation shows that $\Phi - \Phi' = -2\sigma(z_0, z_1)$; therefore, formula (1.16) follows. $\qquad\square$

The action of the Grossmann–Royer operator on tempered distributions can be defined by the formula

$$\langle \widehat{R}(z_0)\psi, \overline{\phi} \rangle = \langle \psi, \overline{\widehat{R}(z_0)\phi} \rangle \tag{1.17}$$

which coincides with the definition

$$(\widehat{R}(z_0)\psi | \phi)_{L^2} = (\psi | \widehat{R}(z_0)\phi)_{L^2}$$

when $\psi$ and $\phi$ are square integrable.

**Example 6.** Let $\psi = \delta_a$ (the Dirac measure centered at $x = a$). We have

$$\langle \widehat{R}(z_0)\delta_a, \overline{\phi} \rangle = \langle \delta_a, \overline{\widehat{R}(z_0)\phi} \rangle = \overline{\widehat{R}(z_0)\phi}(a),$$

i.e.

$$\langle \widehat{R}(z_0)\delta_a, \overline{\phi} \rangle = e^{\frac{2i}{\hbar}p_0 \cdot (x_0 - a)}\overline{\phi(2x_0 - a)}.$$

Since $\overline{\phi(2x_0 - a)} = \langle \delta_{2x_0 - a}, \overline{\phi} \rangle$, we thus have

$$\widehat{R}(z_0)\delta_a = e^{\frac{2i}{\hbar}p_0 \cdot (x_0 - a)}\delta_{2x_0 - a}, \tag{1.18}$$

which we can write, with a slight abuse of notation, and taking the parity of $\delta$ into account,

$$\widehat{R}(z_0)\delta(x - a) = e^{\frac{2i}{\hbar}p_0 \cdot (x_0 - a)}\delta(2x_0 - x - a). \tag{1.19}$$

Proposition 3 carries over without difficulty to Grossmann–Royer operators.

**Proposition 7.** (i) *The mapping* $z_0 \longmapsto \widehat{T}(z_0)$ *is strongly continuous on the space* $\mathcal{S}(\mathbb{R}^n)$, *i.e.*

$$\lim_{z \to z_0} \|\widehat{T}(z)\psi - \widehat{T}(z_0)\psi\|'_{\alpha,\beta} = 0,$$

*for every* $\psi \in \mathcal{S}(\mathbb{R}^n)$ *and* $\alpha, \beta \in \mathbb{N}^n$, *where* $\|\cdot\|'_{\alpha,\beta}$ *is the seminorm* (1.2) *on* $\mathcal{S}(\mathbb{R}^n)$.

(ii) *It is weakly* $*$-*continuous on the space* $\mathcal{S}'(\mathbb{R}^n)$.

**Proof.** It is an immediate consequence of the definition

$$\widehat{R}(z_0) = \widehat{T}(z_0)\widehat{R}(0)\widehat{T}(z_0)^{-1}$$

using Proposition 3.                                                    □

## 1.3.  A Functional Relation Between $\widehat{T}(z_0)$ and $\widehat{R}(z_0)$

Let us show a very important property which shows that the Heisenberg–Weyl and Grossmann–Royer operators are related by the symplectic Fourier transform

$$F_\sigma a(z) = a_\sigma(z) = \left(\frac{1}{2\pi\hbar}\right)^n \int_{\mathbb{R}^{2n}} e^{-\frac{i}{\hbar}\sigma(z,z')} a(z') dz'$$

(see Appendix B for the properties of this variant of the usual Fourier transform on $\mathbb{R}^{2n}$).

**Proposition 8.** *Let* $\psi \in \mathcal{S}'(\mathbb{R}^n)$. *We have*

$$\widehat{R}(z_0)\psi(x) = 2^{-n} F_\sigma[\widehat{T}(\cdot)\psi(x)](-z_0), \tag{1.20}$$

*where* $F_\sigma$ *is the symplectic Fourier transform.*

**Proof.** Since $\widehat{R}(z_0)$ and $F_\sigma$ are continuous automorphisms of $\mathcal{S}'(\mathbb{R}^n)$ it is sufficient to assume that $\psi \in \mathcal{S}(\mathbb{R}^n)$. Formula (1.20) follows from (1.15): using the explicit expressions of $\sigma(z_0, z')$ and $\widehat{T}(z')\psi(x)$, the right-hand side of (1.20) is given by

$$\begin{aligned}
A &= \left(\frac{1}{4\pi\hbar}\right)^n \int_{\mathbb{R}^{2n}} e^{\frac{i}{\hbar}\sigma(z_0,z')} \widehat{T}(z')\psi(x) dz' \\
&= \left(\frac{1}{4\pi\hbar}\right)^n \int_{\mathbb{R}^{2n}} e^{\frac{i}{\hbar}(p_0 \cdot x' - p' \cdot x_0 + p' \cdot x - \frac{1}{2}p' \cdot x')} \psi(x - x') dz' \\
&= \left(\frac{1}{4\pi\hbar}\right)^n \int_{\mathbb{R}^n} \left(\int_{\mathbb{R}^n} e^{\frac{i}{\hbar}p' \cdot (x - x_0 - \frac{1}{2}x')} dp'\right) e^{\frac{i}{\hbar}p_0 \cdot x'} \psi(x) dx'.
\end{aligned}$$

We have, using the Fourier inversion formula,

$$\int_{\mathbb{R}^2} e^{\frac{i}{\hbar} p' \cdot (x - x_0 - \frac{1}{2}x')} dp' = (2\pi\hbar)^n \delta(x - x_0 - \tfrac{1}{2}x'),$$

hence, setting $y = \frac{1}{2}x'$,

$$A = 2^{-n} \int_{\mathbb{R}^n} \delta(x - x_0 - \tfrac{1}{2}x') e^{\frac{i}{\hbar} p_0 \cdot x'} \psi(x) dx'$$

$$= \int_{\mathbb{R}^n} \delta(y + x_0 - x) e^{\frac{2i}{\hbar} p_0 \cdot y} \psi(x) dy$$

$$= e^{\frac{2i}{\hbar} p_0 \cdot (x - x_0)} \psi(-x + 2x_0),$$

which proves (1.15). $\qquad\square$

We have seen that the Grossmann–Royer operators are involutions; more generally, we have the following result which is a generalization of the fact that the product of two reflections is a translation.

**Proposition 9.** *The Grossmann–Royer operators satisfy the product formula*

$$\widehat{R}(z_0)\widehat{R}(z_1) = e^{-\frac{2i}{\hbar}\sigma(z_0, z_1)} \widehat{T}(2(z_0 - z_1)) \qquad (1.21)$$

*for all $z_0, z_1 \in \mathbb{R}^{2n}$.*

**Proof.** We have

$$\widehat{R}(z_0)\widehat{R}(z_1)\psi(x) = \widehat{R}(z_0)[e^{\frac{2i}{\hbar} p_1 \cdot (2x_0 - x - x_1)} \psi(2x_1 - x)]$$

$$= e^{\frac{2i}{\hbar} p_0 \cdot (x - x_0)} e^{\frac{2i}{\hbar} p_1 \cdot (x - x_1)} \psi(2x_1 - (2x_0 - x))$$

$$= e^{\frac{i}{\hbar}\Phi} \psi(x - 2(x_0 - x_1));$$

the phase $\Phi$ being given by

$$\Phi = 2[(p_0 - p_1)x - p_0 x_0 - p_1 x_1 + 2p_1 x_0].$$

On the other hand, we have

$$\widehat{T}(2(z_0 - z_1))\psi(x) = e^{\frac{i}{\hbar}\Phi'} \psi(x - 2(x_0 - x_1))$$

with

$$\Phi' = 2((p_0 - p_1)x - (p_0 - p_1)(x_0 - x_1)).$$

We have $\Phi - \Phi' = -2\sigma(z_0, z_1)$ hence (1.21). $\qquad\square$

## 1.4. Quantization of Exponentials

Consider the elementary exponential functions $e^{\frac{i}{\hbar}x_0 \cdot x}$ and $e^{\frac{i}{\hbar}p_0 \cdot p}$. Recalling that the Heisenberg–Weyl operator

$$\widehat{T}(z_0) = e^{-\frac{i}{\hbar}(\widehat{p} \cdot x_0 - p_0 \cdot \widehat{x})}$$

($z_0 = (x_0, p_0)$) is explicitly defined by

$$T(z_0)\psi(x) = e^{\frac{i}{\hbar}(p_0 \cdot x - \frac{i}{2} p_0 \cdot x_0)}\psi(x - x_0), \tag{1.22}$$

we have, in particular,

$$e^{\frac{i}{\hbar}p_0 \cdot \widehat{x}}\psi(x) = e^{\frac{i}{\hbar}p_0 \cdot x}\psi(x), \quad e^{-\frac{i}{\hbar}\widehat{p} \cdot x_0}\psi(x) = \psi(x - x_0). \tag{1.23}$$

Consider now the operator

$$\widehat{M}(z_0) = e^{\frac{i}{\hbar}(\widehat{x} \cdot x_0 + \widehat{p} \cdot p_0)}$$

sometimes called "Weyl's characteristic function" (it appears in the definition of Weyl operators, as we will see in Chapter 4). We are using here the notation $\widehat{x} = (\widehat{x}_1, \ldots, \widehat{x}_n)$ and $\widehat{p} = (\widehat{p}_1, \ldots, \widehat{p}_n)$ where $\widehat{x}_j$ and $\widehat{p}_j$ are the usual operators "multiplication by $x_j$" and $-i\hbar\partial_{x_j}$. The Heisenberg–Weyl operator is related to $\widehat{M}(z_0)$ by the simple formula $\widehat{T}(-Jz_0) = \widehat{M}(z_0)$ hence

$$\widehat{M}(z_0)\psi(x) = e^{\frac{i}{\hbar}(x_0 \cdot x + \frac{1}{2} p_0 \cdot x_0)}\psi(x + p_0), \tag{1.24}$$

and in particular

$$e^{\frac{i}{\hbar}x_0 \cdot \widehat{x}}\psi(x) = e^{\frac{i}{\hbar}x_0 \cdot x}\psi(x), \quad e^{\frac{i}{\hbar}\widehat{p} \cdot p_0}\psi(x) = \psi(x + p_0). \tag{1.25}$$

The result below is essential because it is the key to the quantization of arbitrary observables. Roughly speaking, a "quantization" is a continuous linear mapping

$$\mathrm{Op} : \mathcal{S}'(\mathbb{R}^{2n}) \longrightarrow \mathcal{L}(\mathcal{S}(\mathbb{R}^n), \mathcal{S}'(\mathbb{R}^n))$$

associating to every symbol (or "generalized observable") $a$ defined on $\mathbb{R}^{2n}$ a continuous operator $\widehat{A} = \mathrm{Op}(a)$ defined on some dense subspace of $L^2(\mathbb{R}^n)$ (here $\mathcal{S}(\mathbb{R}^n)$), and satisfying some additional *ad hoc* conditions: for instance, the symbol $a = 1$ should correspond the identity operator. We will have more to say about quantization later in this book, but here is an interesting general property.

**Proposition 10.** *Let* $X(x_0) = e^{\frac{i}{\hbar} x_0 \cdot x}$ *and* $P(p_0) = e^{\frac{i}{\hbar} p_0 \cdot p}$. *Let* Op *be an arbitrary quantization and set* $\widehat{X}_0 = \mathrm{Op}(X(x_0)$ *and* $\widehat{P}_0 = \mathrm{Op}(P(p_0))$. *We have*

$$\widehat{X}_0 = e^{\frac{i}{\hbar} x_0 \cdot \widehat{x}} \quad and \quad \widehat{P}_0 = e^{\frac{i}{\hbar} p_0 \cdot \widehat{p}}, \tag{1.26}$$

*i.e.*

$$\widehat{X}_0 \psi(x) = e^{\frac{i}{\hbar} x_0 \cdot x} \psi(x), \quad \widehat{P}_0 \psi(x) = \psi(x + p_0). \tag{1.27}$$

*The operators* $\widehat{X}_0$ *and* $\widehat{P}_0$ *satisfy the relations*

$$\widehat{X}_0 \widehat{P}_0 = e^{-\frac{1}{2i\hbar} p_0 \cdot x_0} \widehat{M}(z_0), \quad \widehat{P}_0 \widehat{X}_0 = e^{\frac{1}{2i\hbar} p_0 \cdot x_0} \widehat{M}(z_0),$$

*where* $\widehat{M}(z_0) = e^{\frac{i}{\hbar}(x_0 \cdot \widehat{x} + p_0 \cdot \widehat{p})}$. *Hence*

$$[\widehat{X}_0, \widehat{P}_0] = -2i \sin\left(\frac{p_0 \cdot x_0}{2\hbar}\right) \widehat{M}(z_0).$$

**Proof.** It is sufficient to consider the case $n = 1$. Writing $x = x_1$ and $p = p_1$ and expanding the exponential $e^{\frac{i}{\hbar} x_0 x}$ in a Taylor series, we have, in view of the continuity of Op,

$$\widehat{X}(x_0)\psi(x) = \sum_{k=0}^{\infty} \frac{1}{k!} \left(\frac{i}{\hbar} x_0 \cdot x\right)^k \psi(x) = e^{\frac{i}{\hbar} x_0 \cdot x} \psi(x),$$

and similarly

$$\widehat{P}(p_0)\psi(x) = \sum_{k=0}^{\infty} \frac{1}{k!} \left(\frac{i}{\hbar} p_0 \cdot (-i\hbar \partial_x)\right)^k \psi(x) = \psi(x + p_0).$$

Using the Baker–Campbell–Hausdorff formula

$$e^{A+B} = e^{-\frac{1}{2}[A,B]} e^A e^B = e^{\frac{1}{2}[A,B]} e^B e^A \tag{1.28}$$

valid for all operators $A$ and $B$ commuting with $[A, B]$, we have

$$\widehat{X}_0 \widehat{P}_0 = e^{\frac{i}{\hbar} x_0 \cdot \widehat{x}} e^{\frac{i}{\hbar} p_0 \cdot \widehat{p}} = e^{-\frac{1}{2i\hbar} p_0 \cdot x_0} e^{\frac{i}{\hbar}(x_0 \cdot \widehat{x} + p_0 \cdot \widehat{p})},$$

$$\widehat{P}_0 \widehat{X}_0 = e^{\frac{i}{\hbar} p_0 \cdot \widehat{p}} e^{\frac{i}{\hbar} x_0 \cdot \widehat{x}} = e^{\frac{1}{2i\hbar} p_0 \cdot x_0} e^{\frac{i}{\hbar}(x_0 \cdot \widehat{x} + p_0 \cdot \widehat{p})}. \qquad \square$$

We mention that it is also common in the quantum mechanical literature to use the parametrized families of operators

$$U(\alpha) = e^{\frac{i}{\hbar} \alpha \cdot \widehat{p}} \quad and \quad V(\beta) = e^{i\beta \cdot \widehat{x}}$$

where $\alpha$ and $\beta$ are in $\mathbb{R}^n$. These are simply the operators $\mathrm{Op}(P(p_0)$ and $\mathrm{Op}(X(x_0))$ defined by (1.26) with $p_0$ and $x_0$ replaced with $\alpha$ and $\hbar\beta$. They satisfy the commutation relations

$$U(\alpha)V(\beta) = e^{i\alpha \cdot \beta} V(\beta)U(\alpha).$$

## Main Formulas in Chapter 1

| | |
|---|---|
| Heisenberg–Weyl | $\widehat{T}(z_0)\psi(x) = e^{\frac{i}{\hbar}(p_0 \cdot x - \frac{1}{2}p_0 \cdot x_0)}\psi(x - x_0)$ |
| Product formula for HW | $\widehat{T}(z_0)\widehat{T}(z_1) = e^{\frac{i}{\hbar}\sigma(z_0,z_1)}\widehat{T}(z_1)\widehat{T}(z_0)$ |
| Addition formula for HW | $\widehat{T}(z_0 + z_1) = e^{-\frac{i}{2\hbar}\sigma(z_0,z_1)}\widehat{T}(z_0)\widehat{T}(z_1)$ |
| Inversion of HW | $\widehat{T}(z_0)^{-1} = \widehat{T}(z_0)^* = \widehat{T}(-z_0)$ |
| Grossmann–Royer | $\widehat{R}(z_0) = \widehat{T}(z_0)\widehat{R}(0)\widehat{T}(z_0)^{-1}$ |
| Explicit formula for GR | $\widehat{R}(z_0)\psi(x) = e^{\frac{2i}{\hbar}p_0 \cdot (x - x_0)}\psi(2x_0 - x)$ |
| Product formula for GR | $\widehat{R}(z_0)\widehat{R}(z_1) = e^{-\frac{2i}{\hbar}\sigma(z_0,z_1)}\widehat{T}(2(z_0 - z_1))$ |
| Involution formula for GR | $\widehat{R}(z_0)\widehat{R}(z_0) = I_{\mathrm{d}}$ |
| Symplectic FT | $\widehat{R}(z_0)\psi(x) = 2^{-n}F_\sigma[\widehat{T}(\cdot)\psi(x)](-z_0)$ |

# Chapter 2

# The Cross-Wigner Transform

**Summary 11.** The cross-Wigner transform $W(\psi, \phi)$ of the square integrable functions $\psi$ and $\phi$ is defined in terms of the Grossmann–Royer reflection operator; this definition is equivalent to the textbook definition but has certain conceptual and calculational advantages. The cross-Wigner transform satisfies the Moyal identity and can be extended to tempered distributions. The function $W\psi = W(\psi, \psi)$ is the usual Wigner transform of $\psi$.

## 2.1. Definitions of the Cross-Wigner Transform

We begin by giving two definitions of the (cross-)Wigner transform. The first is very concise, and makes use of the Grossmann–Royer operator; it is, in a sense, more "natural" than Wigner's original definition, which we derive thereafter.

### 2.1.1. First definition

Recall that the Grossmann–Royer operator is defined by

$$\widehat{R}(z_0)\psi(x) = e^{\frac{2i}{\hbar} p_0 \cdot (x - x_0)} \psi(2x_0 - x)$$

for all $\psi \in \mathcal{S}(\mathbb{R}^n)$ (or, more generally $\psi \in \mathcal{S}'(\mathbb{R}^n)$).

**Definition 12.** Let $\psi$ and $\phi$ be in $L^2(\mathbb{R}^n)$. The complex function $W(\psi, \phi)$ defined by

$$W(\psi, \phi)(z) = \left(\frac{1}{\pi\hbar}\right)^n (\widehat{R}(z)\psi | \phi)_{L^2} \tag{2.1}$$

is called the cross-Wigner transform. The function $W\psi = W(\psi, \psi)$ is called the Wigner transform of $\psi$:

$$W\psi(z) = \left(\frac{1}{\pi\hbar}\right)^n (\widehat{R}(z)\psi|\psi)_{L^2}. \tag{2.2}$$

In some texts $W\psi$ is called the "Wigner–Blokhintsev transform"; in harmonic analysis (particularly time-frequency analysis) it is often called the "Wigner–Ville" distribution. It is easy to see that the definition above makes sense: using the Cauchy–Schwarz inequality we have, the operator $\widehat{R}(z)$ being unitary,

$$
\begin{aligned}
|W(\psi, \phi)(z)|^2 &= \left(\frac{1}{\pi\hbar}\right)^{2n} |(\widehat{R}(z)\psi|\phi)_{L^2}|^2 \\
&\leq \left(\frac{1}{\pi\hbar}\right)^{2n} \|\widehat{R}(z)\psi\|_{L^2}\|\phi\|_{L^2} \\
&= \left(\frac{1}{\pi\hbar}\right)^{2n} \|\psi\|_{L^2}\|\phi\|_{L^2};
\end{aligned}
$$

hence the inequality

$$\|W(\psi, \phi)\|_\infty \leq \left(\frac{1}{\pi\hbar}\right)^n \|\psi\|_{L^2}\|\phi\|_{L^2}. \tag{2.3}$$

The definition (2.1) can be rewritten in terms of distribution brackets as

$$W(\psi, \phi)(z) = \left(\frac{1}{\pi\hbar}\right)^n \langle\widehat{R}(z)\psi, \overline{\phi}\rangle. \tag{2.4}$$

### 2.1.2. Wigner's definition

In the literature, one finds most of the time the following definition: the cross-Wigner transform of the pair $(\psi, \phi)$ of elements of $L^2(\mathbb{R}^n)$ is given by

$$W(\psi, \phi)(z) = \left(\frac{1}{2\pi\hbar}\right)^n \int_{\mathbb{R}^n} e^{-\frac{i}{\hbar}p\cdot y}\psi\left(x + \frac{1}{2}y\right)\overline{\phi\left(x - \frac{1}{2}y\right)}dy \tag{2.5}$$

and the Wigner transform of $\psi \in L^2(\mathbb{R}^n)$ is

$$W\psi(z) = \left(\frac{1}{2\pi\hbar}\right)^n \int_{\mathbb{R}^n} e^{-\frac{i}{\hbar}p\cdot y}\psi\left(x + \frac{1}{2}y\right)\overline{\psi\left(x - \frac{1}{2}y\right)}dy. \tag{2.6}$$

Let us show that both definitions coincide. We begin by noting that

$$(\widehat{R}(z_0)\psi|\phi)_{L^2} = \int_{\mathbb{R}^n} e^{\frac{2i}{\hbar}p_0\cdot(x-x_0)}\psi(2x_0 - x)\overline{\phi(x)}dx$$

and hence, setting $y = 2(x_0 - x)$ this is

$$(\widehat{R}(z_0)\psi|\phi)_{L^2} = 2^{-n}\int_{\mathbb{R}^n} e^{-\frac{i}{\hbar}p_0\cdot y}\psi\left(x_0 + \frac{1}{2}y\right)\overline{\phi\left(x_0 - \frac{1}{2}y\right)}dy,$$

which is $= (\pi\hbar)^n W(\psi,\phi)(z_0)$, proving (2.6) in view of (2.1).

### 2.1.3. The Gabor transform and its variants

A mathematical object closely related to the Wigner function is the Gabor transform used in signal theory and time-frequency analysis.

**Definition 13.** The Gabor transform (or windowed Fourier transform) with window $\phi \in \mathcal{S}(\mathbb{R}^n)$ is defined by

$$V_\phi\psi(z) = \int_{\mathbb{R}^n} e^{-2\pi i p\cdot x'}\psi(x')\overline{\phi(x'-x)}dx'. \tag{2.7}$$

In the time-frequency literature, $V_\phi\psi$ is also called the "short-time Fourier transform (STFT)".

We note the following rescaling result, whose (trivial) proof is left to the reader as an exercise.

**Lemma 14.** *For real* $\lambda \neq 0$, *set* $\psi_\lambda(x) = \psi(\lambda x)$. *We have*

$$V_{\phi_\lambda}\psi_\lambda(x,p) = \lambda^{-n}V_\phi\psi(\lambda x, \lambda^{-1}p). \tag{2.8}$$

Taking $\lambda = \sqrt{2\pi\hbar}$ it is easy to see that the windowed Fourier-transform and the cross-Wigner transform are related by the formulae

$$W(\psi,\phi)(z) = \left(\frac{2}{\pi\hbar}\right)^{n/2} e^{\frac{2i}{\hbar}p\cdot x}V_{\phi^\vee_{\sqrt{2\pi\hbar}}}\psi_{\sqrt{2\pi\hbar}}\left(z\sqrt{\frac{2}{\pi\hbar}}\right), \tag{2.9}$$

where $\phi^\vee(x) = \phi(-x)$; equivalently:

$$V_\phi\psi(z) = \left(\frac{2}{\pi\hbar}\right)^{-n/2} e^{-i\pi p\cdot x}W(\psi_{1/\sqrt{2\pi\hbar}}, \phi^\vee_{1/\sqrt{2\pi\hbar}})\left(z\sqrt{\frac{\pi\hbar}{2}}\right). \tag{2.10}$$

In harmonic analysis one usually takes $\hbar = 1/2\pi$, and writes $\omega$ instead of $p$; this leads to the familiar formulas

$$W(\psi,\phi)(z) = 2^n e^{4\pi i \omega \cdot x} V_{\phi^\vee} \psi(2z) \tag{2.11}$$

and

$$V_\phi \psi(z) = 2^{-n} e^{-i\pi\omega \cdot x} W(\psi, \phi^\vee)\left(\frac{1}{2}z\right). \tag{2.12}$$

We mention that in this context the squared modulus

$$|V_\phi \psi(z)|^2 = 2^{-2n} \left| W(\psi, \phi^\vee)\left(\frac{1}{2}z\right) \right|^2$$

is called the "spectrogram".

### 2.1.4. Extension to tempered distributions

Recall (formula (2.4)) that the cross-Wigner transform of $\psi, \phi \in \mathcal{S}(\mathbb{R}^n)$ is defined by

$$W(\psi,\phi)(z) = \left(\frac{1}{\pi\hbar}\right)^n \langle \widehat{R}(z)\psi, \overline{\phi} \rangle,$$

where $\widehat{R}(z)$ is the Grossmann–Royer reflection operator. It is clear that this definition still makes sense if $\psi \in \mathcal{S}'(\mathbb{R}^n)$ since the distributional bracket is a pairing between $\mathcal{S}(\mathbb{R}^n)$ and $\mathcal{S}'(\mathbb{R}^n)$. Similarly, $W(\psi,\phi)$ is also defined if $\phi \in \mathcal{S}'(\mathbb{R}^n)$ (this is obvious by using the conjugation property $W(\psi,\phi) = \overline{W(\phi,\psi)}$).

**Example 15.** Choose $\psi = \delta_a$ ( the Dirac measure concentrated at $a \in \mathbb{R}^n$). In Example 6 (formula (1.18)) we have shown that

$$\widehat{R}(z_0)\delta_a = e^{\frac{2i}{\hbar} p_0 \cdot (x_0 - a)} \delta_{2x_0 - a}.$$

It follows that

$$W(\delta_a, \phi)(z_0) = \left(\frac{1}{\pi\hbar}\right)^n e^{\frac{2i}{\hbar} p_0 \cdot (x_0 - a)} \overline{\phi(2x_0 - a)}.$$

It is not immediately obvious from definition (2.4) (nor is it obvious from the equivalent definition (2.5)) that $W(\psi,\phi)$ can be defined when *both* $\psi$ and $\phi$ are tempered distributions. However, in the example above if we formally replace $\phi$ with, say, $\delta_b$, we get (which the usual abuse of notation)

$$W(\delta_a, \delta_b)(z_0) = \left(\frac{1}{2\pi\hbar}\right)^n e^{\frac{i}{\hbar}p_0 \cdot (b-a)} \delta_b\left(x_0 - \frac{1}{2}(a+b)\right),$$

i.e.

$$W(\delta_a, \delta_b) = \left(\frac{1}{2\pi\hbar}\right)^n e^{\frac{i}{\hbar}p_0 \cdot (b-a)} \left(\delta_{(a+b)/2} \otimes 1\right). \tag{2.13}$$

This *légerdemain* can be rigorously justified [28, 52].

**Proposition 16.** *The cross-Wigner transform* $W : (\psi, \phi) \longmapsto W(\psi, \phi)$ *extends into a sesquilinear mapping*

$$W : \mathcal{S}'(\mathbb{R}^n) \times \mathcal{S}'(\mathbb{R}^n) \longrightarrow \mathcal{S}'(\mathbb{R}^n) \tag{2.14}$$

*defined by*

$$W(\psi, \phi) = \left(\frac{1}{2\pi\hbar}\right)^{n/2} (I_d \otimes F_2) V(\psi \otimes \overline{\phi}), \tag{2.15}$$

*where $F_2$ is the Fourier transform in the set of variables $p$ and $V$ is the coordinate change operator on $\mathbb{R}^n$ defined by*

$$Vf(x, y) = f\left(x + \frac{1}{2}y, x - \frac{1}{2}y\right). \tag{2.16}$$

**Proof.** In view of definition (2.5) of the cross-Wigner transform, we have

$$W(\psi, \phi)(x, p) = \left(\frac{1}{2\pi\hbar}\right)^n \int_{\mathbb{R}^n} e^{-\frac{i}{\hbar}p \cdot y} \psi\left(x + \frac{1}{2}y\right) \overline{\phi\left(x - \frac{1}{2}y\right)} dy$$

$$= \left(\frac{1}{2\pi\hbar}\right)^n \int_{\mathbb{R}^n} e^{-\frac{i}{\hbar}p \cdot y} V(\psi \otimes \overline{\phi})(x, y) dy,$$

which is (2.15) when $\psi, \phi \in \mathcal{S}(\mathbb{R}^n)$. Now, the operator $V$ obviously maps $\mathcal{S}'(\mathbb{R}^n) \otimes \mathcal{S}'(\mathbb{R}^n)$ onto $\mathcal{S}'(\mathbb{R}^{2n})$ and $I_d \otimes F_2$ is an automorphism of $\mathcal{S}'(\mathbb{R}^{2n})$; the result follows. $\square$

We leave it to the reader as an exercise to verify that formula (2.15) yields the same result (2.13) for $W(\delta_a, \delta_b)$ as obtained by the "heuristic" derivation above. In particular, the Wigner transform of the Dirac measure $\delta$ is

$$W\delta = \left(\frac{1}{2\pi\hbar}\right)^n (\delta \otimes 1). \tag{2.17}$$

## 2.2. Properties of the Cross-Wigner Transform

We prove in this section the basic algebraic properties of the cross-Wigner transform.

### 2.2.1. Elementary algebraic properties

We will see (Proposition 18) that $(\psi, \phi) \longmapsto W(\psi, \phi)$ is a sesquilinear mapping $L^2(\mathbb{R}^n) \times L^2(\mathbb{R}^n) \longrightarrow L^2(\mathbb{R}^{2n})$; the sesquilinearity means that

$$W(\psi, \phi_1 + \phi_2) = W(\psi, \phi_1) + W(\psi, \phi_2),$$

$$W(\psi_1 + \psi_2, \phi) = W(\psi_1, \phi) + W(\psi_2, \phi),$$

$$W(\lambda\psi, \phi) = \lambda W(\psi, \phi),$$

$$W(\psi, \lambda\phi) = \overline{\lambda} W(\psi, \phi)$$

($\lambda$ an arbitrary complex number); it follows that we have the so-called *polarization identity*

$$\mathrm{Re}\, W(\psi, \phi) = \frac{1}{4} \left[ W(\psi + \phi) - W(\psi - \phi) \right]. \tag{2.18}$$

The Wigner transform is not additive: we have in general $W(\psi_1 + \psi_2) \neq W\psi_1 + W\psi_2$: this is due to the presence of cross-terms, which are responsible for the occurrence of interferences.

**Proposition 17.** *Let $\psi = \sum_{1 \leq j \leq m} \lambda_j \psi_j$ be a finite linear combination of elements of $L^2(\mathbb{R}^n)$. The Wigner transform $W\psi$ is given by*

$$W\psi = \sum_{j=1}^{m} |\lambda_j|^2 W\psi_j + 2\,\mathrm{Re} \sum_{\substack{k=1 \\ k > \ell}}^{m} \sum_{\ell=1}^{m} \lambda_k \bar{\lambda}_\ell W(\psi_k, \psi_\ell). \tag{2.19}$$

Formula (2.19) is easily proven using an induction on the number $m$ of terms; we leave the details to the reader as an exercise.

The following conjugation property

$$W(\psi, \phi) = \overline{W(\phi, \psi)} \tag{2.20}$$

immediately follows from the integral definition (2.5) of the cross-Wigner transform:

$$\overline{W(\psi, \phi)(z)} = \left(\frac{1}{2\pi\hbar}\right)^n \int_{\mathbb{R}^n} e^{\frac{i}{\hbar}p\cdot y} \overline{\psi\left(x + \frac{1}{2}y\right)\phi\left(x - \frac{1}{2}y\right)} \, dy$$

$$= \left(\frac{1}{2\pi\hbar}\right)^n \int_{\mathbb{R}^n} e^{-\frac{i}{\hbar}p\cdot y'} \overline{\psi\left(x - \frac{1}{2}y'\right)} \phi\left(x + \frac{1}{2}y'\right) \, dy'$$

$$= \overline{W(\phi, \psi)}(z).$$

Hence, in particular *the Wigner transform $W\psi$ is always a real function.* It is however not in general positive. To see this it suffices to choose an odd function $\psi \in \mathcal{S}(\mathbb{R}^n)$; then $\widehat{R}(0)\psi = -\psi$ hence by (2.2)

$$W\psi(0) = -\left(\frac{1}{\pi\hbar}\right)^n \|\psi\|_{L^2} < 0.$$

It follows, by continuity, that there exists $\varepsilon > 0$ such that $W\psi(z) < 0$ for $|z| < \varepsilon$. In fact, a classical result of Hudson [59], extended to the multidimensional case by Soto and Claverie [77], tells us that $W\psi$ is non-negative if and only if $\psi$ is a generalized Gaussian; we will come back to this property in Chapter 9. (We also mention that Janssen [60] has extended Hudson's result.)

The following tensor product properties hold when one splits the $x$ and $p$ variables: if $x = (x', x'') \in \mathbb{R}^k \times \mathbb{R}^{n-k}$, $p = (p', p'') \in \mathbb{R}^k \times \mathbb{R}^{n-k}$ and $\psi' \in L^2(\mathbb{R}^k)$, $\psi'' \in L^2(\mathbb{R}^{n-k})$, then

$$W(\psi' \otimes \psi'') = W'\psi' \otimes W''\psi'', \tag{2.21}$$

where $W'$ and $W''$ are the cross-Wigner transforms on $L^2(\mathbb{R}^k)$ and $L^2(\mathbb{R}^{n-k})$, in that order. More generally, we have

$$W(\psi' \otimes \psi'', \phi' \otimes \phi'') = W'(\psi', \phi') \otimes W''(\psi'', \phi''). \tag{2.22}$$

The proofs of these equalities are elementary, using the first definition of $W(\psi, \phi)$ and noting that the Grossmann–Royer operator splits as

$$\widehat{R}(z', z'') = \widehat{R}(z') \otimes \widehat{R}(z'')$$

for $z = (z', z'')$.

### 2.2.2. Analytical properties and continuity

We begin by showing that a normalization result when $W\psi$ absolutely integrable if $\psi$ is square integrable. This result is important for the study of the probabilistic properties of the Wigner transform.

**Proposition 18.** *Let $\psi, \phi \in L^2(\mathbb{R}^n)$ and suppose that $W(\psi, \phi) \in L^1(\mathbb{R}^{2n})$. Then*

$$\int_{\mathbb{R}^{2n}} W(\psi, \phi)(z_0)dz_0 = (\psi|\phi)_{L^2}. \qquad (2.23)$$

*In particular, if $\psi \in L^2(\mathbb{R}^n)$ and $W\psi \in L^1(\mathbb{R}^{2n})$ then $W\psi \in L^1(\mathbb{R}^{2n})$ and*

$$\int_{\mathbb{R}^{2n}} W\psi(z_0)dz_0 = ||\psi||_{L^2}^2. \qquad (2.24)$$

**Proof.** Assume that $\psi, \phi \in \mathcal{S}(\mathbb{R}^n)$. We begin by noting that by definition of the Grossmann–Royer operator $\widehat{R}(z_0)$ we have

$$\int_{\mathbb{R}^{2n}} \widehat{R}(z_0)\psi(x)dz_0 = \int_{\mathbb{R}^{2n}} e^{\frac{2i}{\hbar}p_0\cdot(x-x_0)}\psi(2x_0 - x)dp_0dx_0$$

$$= \int_{\mathbb{R}^n} e^{\frac{2i}{\hbar}p_0\cdot(x-x_0)}\left(\int_{\mathbb{R}^n} \psi(2x_0 - x)dx_0\right)dp_0$$

$$= (\pi\hbar)^n \int_{\mathbb{R}^n} \delta(x - x_0)\psi(2x_0 - x)dx_0$$

$$= (\pi\hbar)^n \psi(x).$$

It follows that

$$\int_{\mathbb{R}^{2n}} W(\psi, \phi)(z_0)dz_0 = \left(\frac{1}{\pi\hbar}\right)^n \left(\int_{\mathbb{R}^{2n}} \widehat{R}(z_0)\psi(x)\overline{\phi(x)}dz_0\right)dx$$

$$= \left(\frac{1}{\pi\hbar}\right)^n \int_{\mathbb{R}^n} \left(\int_{\mathbb{R}^{2n}} \widehat{R}(z_0)\psi(x)dz_0\right)\overline{\phi(x)}dx$$

$$= \int_{\mathbb{R}^n} \psi(x)\overline{\phi(x)}dx$$

which is formula (2.23). $\qquad \square$

An immediate consequence of the definition of the cross-Wigner transform is that $W(\psi, \phi)$ is bounded and continuous if $\psi$ and $\phi$ are both square integrable.

**Proposition 19.** *Let $\psi$ and $\phi$ be in $L^2(\mathbb{R}^n)$. The function $z \longmapsto W(\psi,\phi)(z)$ is bounded and continuous on $\mathbb{R}^{2n}$, i.e.*

$$W(\psi,\phi) \in L^\infty(\mathbb{R}^{2n}) \cap C^0(\mathbb{R}^{2n}).$$

**Proof.** Let us prove the continuity of $W(\psi,\phi)$. It suffices to do so for $\psi,\phi \in \mathcal{S}(\mathbb{R}^n)$ since $\mathcal{S}(\mathbb{R}^n)$ is dense in $L^2(\mathbb{R}^n)$. Set

$$f(z,u) = \left(\frac{1}{\pi\hbar}\right)^n R(z)\psi(u)\overline{\phi(u)};$$

the function $z \longmapsto f(z,u)$ is continuous and we have $|f(z,u)| \leq g(u)$ where $g(u) \leq C|\phi(u)|$ for some constant $C > 0$ (because $R(z)\psi(u)$ is bounded). The continuity of $W(\psi,\phi)$ now follows applying Lebesgue's dominated convergence theorem. $\qquad\square$

We will show in Section 6 that $W(\psi,\phi)$ is in addition itself square-integrable; in particular we will find that

$$||W\psi||_{L^2(\mathbb{R}^{2n})} = \left(\frac{1}{2\pi\hbar}\right)^{n/2} ||\psi||_{L^2}.$$

Talking about continuity, we have the following more general result.

**Proposition 20.** *Let $\psi,\phi \in \mathcal{S}'(\mathbb{R}^n)$. The function $z \longmapsto W(\psi,\phi)(z)$ is continuous on $\mathbb{R}^{2n}$.*

**Proof.** Assume first that $\psi,\phi \in \mathcal{S}(\mathbb{R}^n)$. Let us show that

$$\lim_{z\to z_0} (W(\psi,\phi)(z) - W(\psi,\phi)(z_0)) = 0. \qquad (2.25)$$

We have shown in Chapter 1 (Proposition 7) that the mapping $z \longmapsto \widehat{R}(z)\psi$ is strongly continuous on $\mathbb{R}^{2n}$; thus we have in particular

$$\lim_{z\to z_0} ||\widehat{R}(z)\psi - \widehat{R}(z_0)\psi||_\infty = 0.$$

In view of the first definition of $W(\psi,\phi)(z)$ we have

$$W(\psi,\phi)(z) - W(\psi,\phi)(z_0) = \left(\frac{1}{\pi\hbar}\right)^n \int_{\mathbb{R}^n} (\widehat{R}(z)\psi(x') - \widehat{R}(z_0)\psi(x'))\overline{\phi(x')}dx';$$

hence

$$|W(\psi,\phi)(z) - W(\psi,\phi)(z_0)| \le \left(\frac{1}{\pi\hbar}\right)^n ||\widehat{R}(z)\psi - \widehat{R}(z_0)\psi||_\infty ||\phi||_1$$

and (2.25). The case $\psi, \phi \in \mathcal{S}'(\mathbb{R}^n)$ follows by duality. $\qquad\square$

### 2.2.3. The marginal properties

Here is a property which is at the origin of the keen interest quantum physicists immediately showed for the Wigner transform.

**Proposition 21.** *Assume that $\psi, \phi \in L^1(\mathbb{R}^n) \cap L^2(\mathbb{R}^n)$. Then*

$$\int_{\mathbb{R}^n} W(\psi,\phi)(z)dp = \psi(x)\overline{\phi(x)}, \qquad (2.26)$$

$$\int_{\mathbb{R}^n} W(\psi,\phi)(z)dx = F\psi(p)\overline{F\phi(p)}; \qquad (2.27)$$

*hence, in particular*

$$\int_{\mathbb{R}^n} W\psi(z)dp = |\psi(x)|^2, \quad \int_{\mathbb{R}^n} W\psi(z)dx = |F\psi(p)|^2. \qquad (2.28)$$

**Proof.** Writing the Fourier inversion formula as

$$\int_{\mathbb{R}^n} e^{-\frac{i}{\hbar}p\cdot y}dp = (2\pi\hbar)^n \delta(y),$$

we have

$$\int_{\mathbb{R}^n} W(\psi,\phi)(z)dp = \int_{\mathbb{R}^n} \delta(y)\psi\left(x + \frac{1}{2}y\right)\overline{\phi\left(x - \frac{1}{2}y\right)}dy$$

$$= \int_{\mathbb{R}^n} \delta(y)\psi(x)\overline{\phi(x)}dy$$

$$= \psi(x)\overline{\phi(x)}$$

as claimed. This proves formula (2.26). Let us prove (2.27). Making the unitary change of variables $(x, y) \longmapsto (x + \frac{1}{2}y, x - \frac{1}{2}y)$ in the right-hand side of the equality

$$\int_{\mathbb{R}^n} W\psi(z)dx = \left(\frac{1}{2\pi\hbar}\right)^n \int_{\mathbb{R}^{2n}} e^{-\frac{i}{\hbar}p\cdot y}\psi\left(x+\frac{1}{2}y\right)\overline{\phi\left(x-\frac{1}{2}y\right)}dxdy,$$

we get, using Fubini's theorem,

$$\int_{\mathbb{R}^n} W\psi(z)dx = \left(\frac{1}{2\pi\hbar}\right)^n \int_{\mathbb{R}^{2n}} e^{-\frac{i}{\hbar}p\cdot x}\psi(x')\overline{e^{-\frac{i}{\hbar}p\cdot x}\phi(x)}dxdy$$

$$= \left(\frac{1}{2\pi\hbar}\right)^n \int_{\mathbb{R}^n} e^{-\frac{i}{\hbar}p\cdot x}\psi(x)dx\overline{\int_{\mathbb{R}^n} e^{-\frac{i}{\hbar}p\cdot y}\phi(y)dy}$$

$$= F\psi(p)\overline{F\phi(p)},$$

which is formula (2.27). The two formulas (2.28) trivially follow by taking $\psi = \phi$. □

Notice that by integrating with respect to $x$ (respectively, $p$) the first (respectively, second) formula (2.28), we recover the relation

$$\int_{\mathbb{R}^{2n}} W\psi(z)dz = ||\psi||_{L^2}.$$

### 2.2.4. Translating Wigner transforms

The following important result describes the behavior of the cross-Wigner transform under translations.

**Proposition 22.** *For every pair $(\psi, \phi)$ of functions in $\mathcal{S}'(\mathbb{R}^n)$, we have*

$$W(\widehat{T}(z_0)\psi, \widehat{T}(z_1)\phi)(z) = e^{-\frac{i}{\hbar}[\sigma(z,z_0-z_1)+\frac{1}{2}\sigma(z_0,z_1)]}W(\psi,\phi)(z-\langle z\rangle),$$
(2.29)

*where $\langle z\rangle = \frac{1}{2}(z_0 + z_1)$.*

**Proof.** To prove formula (2.29) it is sufficient to assume that $\psi, \phi \in \mathcal{S}(\mathbb{R}^n)$ the general case following by duality. We will set $\langle x\rangle = \frac{1}{2}(x_0 + x_1)$ and $\langle p\rangle = \frac{1}{2}(p_0 + p_1)$. By definition of the Heisenberg–Weyl operator, we have

$$\widehat{T}(z_0)\psi\left(x+\frac{1}{2}y\right) = e^{\frac{i}{\hbar}[p_0\cdot(x+\frac{1}{2}y)-\frac{1}{2}p_0\cdot x_0]}\psi\left(x-x_0+\frac{1}{2}y\right),$$

$$\widehat{T}(z_1)\phi\left(x-\frac{1}{2}y\right) = e^{\frac{i}{\hbar}[p_1\cdot(x-\frac{1}{2}y)-\frac{1}{2}p_1\cdot x_1]}\phi\left(x-x_1-\frac{1}{2}y\right),$$

and hence

$$\widehat{T}(z_0)\psi\left(x + \frac{1}{2}y\right)\overline{\widehat{T}(z_1)\phi\left(x - \frac{1}{2}y\right)} = e^{\frac{i}{\hbar}\delta(z_0, z_1)}e^{\frac{i}{\hbar}\langle p\rangle \cdot y}\psi\left(x - x_0 + \frac{1}{2}y\right)$$

$$\times \overline{\phi\left(x - x_1 - \frac{1}{2}y\right)}$$

with

$$\delta(z_0, z_1) = (p_0 - p_1) \cdot x - \frac{1}{2}(p_0 \cdot x_0 - p_1 \cdot x_1).$$

It follows that we have

$$W(\widehat{T}(z_0)\psi, \widehat{T}(z_1)\phi)(z) = \left(\frac{1}{2\pi\hbar}\right)^n e^{\frac{i}{\hbar}\delta(z_0, z_1)}\int_{\mathbb{R}^n} e^{-\frac{i}{\hbar}(p - \langle p\rangle)\cdot y}\psi\left(x - x_0\right.$$

$$\left. + \frac{1}{2}y\right)\overline{\phi\left(x - x_0 - \frac{1}{2}y\right)}dy.$$

Performing the change of variables $y' = x_1 - x_0 + y$ in the integral this equality becomes

$$W(\widehat{T}(z_0)\psi, \widehat{T}(z_1)\phi)(z) = \left(\frac{1}{2\pi\hbar}\right)^n e^{\frac{i}{\hbar}\Delta}\int_{\mathbb{R}^n} e^{-\frac{i}{\hbar}(p - \langle p\rangle)\cdot y}\psi\left(x - \langle x\rangle\right.$$

$$\left. + \frac{1}{2}y\right)\overline{\phi\left(x - \langle x\rangle - \frac{1}{2}y\right)}dy$$

where the phase $\Delta$ is given by

$$\Delta = (p_0 - p_1) \cdot x - (x_0 - x_1) \cdot p + \frac{1}{2}(p_1 \cdot x_0 - p_0 \cdot x_1)$$

$$= -\sigma(z, z_0 - z_1) - \frac{1}{2}\sigma(z_0, z_1),$$

hence formula (2.29). $\qquad\square$

**Corollary 23.** *We have the following translation formulas:*

$$W(\widehat{T}(z_0)\psi, \widehat{T}(z_0)\phi)(z) = W(\psi, \phi)(z - z_0), \qquad (2.30)$$

$$W(\widehat{T}(z_0)\psi)(z) = W\psi(z - z_0), \qquad (2.31)$$

$$W(\widehat{T}(z_0)\psi, \phi)(z) = e^{-\frac{i}{\hbar}\sigma(z, z_0)}W(\psi, \phi)\left(z - \frac{1}{2}z_0\right). \qquad (2.32)$$

**Proof.** The formulas above immediately follow from (2.29). $\qquad\square$

**Example 24.** In quantum mechanics cat states are common in the study of entanglement (the terminology comes from the "Schrödinger cat"). These states are of the type $\phi = \widehat{T}(z_0)\psi + \widehat{T}(-z_0)\psi$. Using the formulas above, the Wigner transform of $\phi$ is

$$W\phi(z) = W\psi(z - z_0) + W\psi(z + z_0) + 2\,\operatorname{Re}(e^{\frac{i}{\hbar}\sigma(z,z_0)}W\psi(z)).$$

The term $e^{\frac{i}{\hbar}\sigma(z,z_0)}W\psi(z)$ is at the origin of "sub-Planckian" interference effects.

## Main Formulas in Chapter 2

| | |
|---|---|
| CWT: definition 1 | $W(\psi, \phi)(z) = \left(\frac{1}{\pi\hbar}\right)^n (\widehat{R}(z)\psi|\phi)_{L^2})$ |
| CWT: definition 2 | $W(\psi, \phi)(z) = \left(\frac{1}{2\pi\hbar}\right)^n$ |
| | $\int_{\mathbb{R}^n} e^{-\frac{i}{\hbar}p\cdot y}\psi(x + \frac{1}{2}y)\overline{\phi(x - \frac{1}{2}y)}dy$ |
| Wigner | $W\psi(z) = \left(\frac{1}{2\pi\hbar}\right)^n \int_{\mathbb{R}^n} e^{-\frac{i}{\hbar}p\cdot y}\psi(x + \frac{1}{2}y)\overline{\psi(x - \frac{1}{2}y)}dy$ |
| STFT | $V_\phi\psi(z) = \int_{\mathbb{R}^n} e^{-2\pi i p\cdot x'}\psi(x')\overline{\phi(x' - x)}dx'$ |
| STFT–CWT | $V_\phi\psi(z) = 2^{-n}e^{-i\pi\omega\cdot x}W(\psi, \phi^\vee)(\frac{1}{2}z)$ |
| Normalization | $\|W\psi\|_{L^1(\mathbb{R}^{2n})} = \|\psi\|_{L^2}$ |
| Boundedness | $\|W(\psi, \phi)\|_\infty \le \left(\frac{1}{\pi\hbar}\right)^n \|\psi\|_{L^2}\|\phi\|_{L^2}$ |
| $L^2$-boundedness | $\|W\psi\|_{L^2(\mathbb{R}^{2n})} = \left(\frac{1}{2\pi\hbar}\right)^{n/2} \|\psi\|_{L^2}$ |
| Translation | $W(\widehat{T}(z_0)\psi, \widehat{T}(z_0)\phi)(z) = W(\psi, \phi)(z - z_0)$ |

# Chapter 3

# The Cross-Ambiguity Function

**Summary 25.** The cross-ambiguity function is defined concisely in terms of the Heisenberg–Weyl operator. It is the symplectic Fourier transform of the cross-Wigner transform, and has algebraic and analytical properties very similar to those of the latter.

We defined previously the cross-Wigner transform using the Heisenberg–Weyl shift operator. In the present chapter we define the cross-ambiguity function using the Grossmann–Royer parity operator; we will see that it is essentially the Fourier transform of the cross-Wigner transform.

## 3.1. Definition of the Cross-Ambiguity Function

### 3.1.1. Definition using the Heisenberg–Weyl operator

Recall that the Heisenberg–Weyl operator $\widehat{T}(z_0)$ is defined by

$$\widehat{T}(z_0)\psi = e^{\frac{i}{\hbar}(p_0 \cdot x - \frac{1}{2} p_0 \cdot x_0)} \psi_0(x - x_0),$$

where $\psi \in \mathcal{S}(\mathbb{R}^n)$ (or $\psi \in \mathcal{S}'(\mathbb{R}^n)$).

**Definition 26.** Let $\psi$ and $\phi$ be in $L^2(\mathbb{R}^n)$. The cross-ambiguity function of the pair $(\psi, \phi)$ is defined by

$$A(\psi, \phi)(z) = \left( \frac{1}{2\pi\hbar} \right)^n (\psi | \widehat{T}(z)\phi)_{L^2}. \tag{3.1}$$

The function $A\psi = A(\psi, \psi)$ is called the (auto-) ambiguity function:

$$A\psi(z) = \left( \frac{1}{2\pi\hbar} \right)^n (\psi | \widehat{T}(z)\psi)_{L^2}. \tag{3.2}$$

Notice that $A\psi$ is not in general a real function; in fact $\overline{A\psi} = A(\psi^\vee)$ where $\psi^\vee(x) = \psi(-x)$, so $A\psi$ is real only if $\psi$ is even. More generally, we have

$$\overline{A(\psi,\phi)} = A(\phi^\vee, \psi^\vee), \tag{3.3}$$

which easily follows from the equality

$$\overline{A(\psi,\phi)(z)} = \left(\frac{1}{2\pi\hbar}\right)^n \overline{(\phi|\widehat{T}(z)\psi)_{L^2}}. \tag{3.4}$$

As in the case of the cross-Wigner transform, formula (3.2) defines a bounded function on $L^2(\mathbb{R}^n) \times L^2(\mathbb{R}^n)$: using the definition (3.4) of $A(\psi,\phi)$ and the Cauchy–Schwarz inequality, we have

$$\begin{aligned}
|A(\psi,\phi)(z)| &= \left(\frac{1}{2\pi\hbar}\right)^n |(\psi|\widehat{T}(z)\phi)_{L^2}| \\
&\leq \left(\frac{1}{2\pi\hbar}\right)^n \|\psi\|_{L^2} \|\widehat{T}(z)\phi\|_{L^2} \\
&= \left(\frac{1}{2\pi\hbar}\right)^n \|\psi\|_{L^2} \|\phi\|_{L^2}
\end{aligned}$$

the last equality because $\widehat{T}(z)$ is unitary; taking the supremum with respect to $z$, we get

$$\|A(\psi,\phi)\|_\infty \leq \|\psi\|_{L^2} \|\phi\|_{L^2}. \tag{3.5}$$

When $\hbar = 1/2\pi$ and $p$ is viewed as a frequency, the function $A\psi$ is also called "radar ambiguity function" or "Woodward ambiguity function".

Having in mind our applications to the case where $\psi$ might be a tempered distribution, we notice that definition (3.1) can be written in terms of distributional brackets as

$$A(\psi,\phi)(-z) = \left(\frac{1}{2\pi\hbar}\right)^n \langle \widehat{T}(z)\psi, \overline{\phi} \rangle. \tag{3.6}$$

This formula can be used to define $A(\psi,\phi)$ when $\psi \in \mathcal{S}'(\mathbb{R}^n)$ and $\phi \in \mathcal{S}(\mathbb{R}^n)$.

### 3.1.2. Traditional definition

The definition we have given is not the one finds in the traditional literature, where the cross-ambiguity function is defined in integral form.

**Proposition 27.** *The cross-ambiguity function of* $\psi, \phi \in L^2(\mathbb{R}^n)$ *is explicitly given by the formula*

$$A(\psi, \phi)(z) = \left(\frac{1}{2\pi\hbar}\right)^n \int_{\mathbb{R}^n} e^{-\frac{i}{\hbar}p\cdot y} \psi\left(y + \frac{1}{2}x\right) \overline{\phi\left(y - \frac{1}{2}x\right)} dy. \qquad (3.7)$$

**Proof.** By definition of $\widehat{T}(z)$, we have, setting $z = (x, p)$,

$$A(\psi, \phi)(z) = \left(\frac{1}{2\pi\hbar}\right)^n \int_{\mathbb{R}^n} e^{\frac{i}{\hbar}(p\cdot x'' - \frac{1}{2}p\cdot x)} \psi^\vee(x'' - x)\overline{\phi^\vee(x'')} dx'', \qquad (3.8)$$

which is precisely (3.8) performing the change of variables $x'' = -y + \frac{1}{2}x$. $\qquad \square$

In particular, the ambiguity function is thus given by

$$A\psi(z) = \left(\frac{1}{2\pi\hbar}\right)^n \int_{\mathbb{R}^n} e^{-\frac{i}{\hbar}p\cdot y} \psi\left(y + \frac{1}{2}x\right) \overline{\psi\left(y - \frac{1}{2}x\right)} dy. \qquad (3.9)$$

### 3.1.3. The Fourier–Wigner transform

In some texts (e.g. [28]) one introduces the transform $V : (f, g) \longmapsto V(f, g)$ on $L^2(\mathbb{R}^n) \times L^2(\mathbb{R}^n)$ defined by

$$V(f, g)(p, x) = \int_{-\infty}^{\infty} e^{2\pi i x y} f\left(y + \frac{1}{2}p\right) \overline{g\left(y - \frac{1}{2}p\right)} dy. \qquad (3.10)$$

It is usually called the "Fourier–Wigner transform". Rewriting this formula as

$$V(f, g)(x, -p) = \int_{-\infty}^{\infty} e^{-2\pi i p y} f\left(y + \frac{1}{2}x\right) \overline{g\left(y - \frac{1}{2}x\right)} dy$$

comparison with the integral expression (3.7) of the cross-ambiguity function shows that we have

$$V(f, g)(x, -p) = A(f, g)(x, p)$$

when one chooses $\hbar = 1/2\pi$; rewriting this formula in the form

$$V(f, g)(p, x) = A(f, g)(p, -x) = A(f, g)(J(x, p))$$

suggests the following $\hbar$-dependent redefinition of the Fourier–Wigner transform.

**Definition 28.** The Fourier–Wigner transform of $\psi, \phi \in L^2(\mathbb{R}^n)$ is defined, for $z = (x, p) \in \mathbb{R}^{2n}$, by

$$V(\psi, \phi)(p, x) = A(\psi, \phi)(p, -x).$$

It is explicitly given by

$$V(\psi, \phi)(p, x) = \int_{\mathbb{R}^n} e^{\frac{i}{\hbar} xy} \psi\left(y + \frac{1}{2}p\right) \overline{\phi\left(y - \frac{1}{2}p\right)} dy. \tag{3.11}$$

The properties of the Fourier–Wigner transform are trivially obtained from those of the cross-ambiguity function since they are related by a linear change of variables.

## 3.2. Properties and Relation with the Wigner Transform

### 3.2.1. Properties of the cross-ambiguity function

We have, as for the cross-Wigner transform, a polarization identity:

$$Re A(\psi, \phi) = \frac{1}{4}\left[A(\psi + \phi) - A(\psi - \phi)\right]; \tag{3.12}$$

the latter immediately follows from the fact that the mapping $(\psi, \phi) \longmapsto A(\psi, \phi)$ is sesquilinear.

As the Wigner transform, the ambiguity function behaves well under tensor products: if $x = (x', x'')$ with $x' \in \mathbb{R}^k$, $x'' \in \mathbb{R}^{n-k}$ and $\psi' \in \mathcal{S}(\mathbb{R}^k)$, $\psi'' \in \mathcal{S}(\mathbb{R}^{n-k})$, then

$$A(\psi' \otimes \psi'') = A'\psi' \otimes A''\psi'' \tag{3.13}$$

and

$$A(\psi' \otimes \psi'', \phi' \otimes \phi'') = A'(\psi', \phi') \otimes A''(\psi'', \phi''). \tag{3.14}$$

Here are some elementary continuity properties of the cross-ambiguity function; they are quite similar to those of the cross-Wigner transform.

**Proposition 29.** *The cross-ambiguity function has the following properties:*

(i) *It is a continuous mapping $\mathcal{S}(\mathbb{R}^n) \times \mathcal{S}(\mathbb{R}^n) \longrightarrow \mathcal{S}(\mathbb{R}^{2n})$.*

(ii) *That mapping extends into a continuous mapping*

$$A : L^2(\mathbb{R}^n) \times L^2(\mathbb{R}^n) \longrightarrow C^0(\mathbb{R}^{2n}) \cap L^\infty(\mathbb{R}^{2n}).$$

**Proof.** (i) In view of formula (3.8) and the fact that multiplication by the exponential $e^{-ip\cdot x/2\hbar}$ is an automorphism of $S(\mathbb{R}^{2n})$, it suffices to show that the function $F$ defined by

$$F(z) = \int_{\mathbb{R}^n} e^{\frac{i}{\hbar} p\cdot x'} \psi(x' - x)\overline{\phi(x')}dx'$$

is in $S(\mathbb{R}^{2n})$ if $\psi$ and $\phi$ are in $S(\mathbb{R}^n)$. Now, $F$ is, up to a factor, the partial Fourier transform in $x'$ of the mapping $(x, x') \longmapsto f(x, x') = \psi(x' - x)\overline{\phi(x')}$; since $f \in S(\mathbb{R}^{2n})$ the claim follows. The fact that $A(\psi, \phi)$ is continuous follows, since $S(\mathbb{R}^n)$ is dense in $L^2(\mathbb{R}^n)$. $\qquad\square$

**Proposition 30.** *Let* $\psi, \phi \in S(\mathbb{R}^n)$. *The mapping* $z \longmapsto A(\psi, \phi)(z)$ *is continuous on* $\mathbb{R}^{2n}$.

**Proof.** The proof is *mutatis mutandis* the same as that of Proposition 20 in Chapter 2. We have shown in Chapter 1 (Proposition 3) that the mapping $z \longmapsto \widehat{T}(z)\psi$ is strongly continuous on $\mathbb{R}^{2n}$; in particular

$$\lim_{z \to z_0} ||\widehat{T}(z)\psi - \widehat{T}(z_0)\psi||_\infty = 0.$$

By definition of the cross-ambiguity function in terms of the Heisenberg–Weyl operator, we have

$$A(\psi, \phi)(z) - A(\psi, \phi)(z_0) = \left(\frac{1}{2\pi\hbar}\right)^n \int_{\mathbb{R}^n} (\widehat{T}(z)\psi(x') - \widehat{T}(z_0)\psi(x'))\overline{\phi(x')}dx'$$

and hence

$$|W(\psi, \phi)(z) - W(\psi, \phi)(z_0)| \le \left(\frac{1}{\pi\hbar}\right)^n ||\widehat{R}(z)\psi - \widehat{R}(z_0)\psi||_\infty ||\phi||_1.$$

The result follows as in the proof of Proposition 3. $\qquad\square$

### 3.2.2. Relation with the cross-Wigner transform

We begin by showing that the cross-ambiguity function and the cross-Wigner transform are related by a symplectic Fourier transform. This is a very important result, since it allows to easily toggle between both objects.

**Proposition 31.** *Let* $\psi$ *and* $\phi$ *be in* $S'(\mathbb{R}^n)$. *We have*

$$A(\psi, \phi) = F_\sigma W(\psi, \phi), \quad W(\psi, \phi) = F_\sigma A(\psi, \phi). \qquad (3.15)$$

*In particular* $A\psi = F_\sigma(W\psi)$.

**Proof.** The two equations in (3.15) are equivalent since the symplectic Fourier transform $F_\sigma$ is an involution of $\mathcal{S}'(\mathbb{R}^{2n})$. It is therefore sufficient to show that $A(\psi, \phi) = F_\sigma W(\psi, \phi)$; by duality, it is moreover sufficient to assume that $\psi$ and $\phi$ are in $\mathcal{S}(\mathbb{R}^n)$. Setting

$$A = (2\pi\hbar)^{2n} F_\sigma W(\psi, \phi),$$

we have, by definition of $F_\sigma$ and $W(\psi, \phi)$,

$$A(z) = \int_{\mathbb{R}^{3n}} e^{-\frac{i}{\hbar}[\sigma(z,z') + p' \cdot (y-x)]} \psi\left(x' + \frac{1}{2}y\right) \overline{\phi\left(x' - \frac{1}{2}y\right)} dp' dx' dy$$

$$= \int_{\mathbb{R}^{3n}} e^{-\frac{i}{\hbar} p' \cdot (y-x)} e^{-\frac{i}{\hbar} p \cdot x'} \psi\left(x' + \frac{1}{2}y\right) \overline{\phi\left(x' - \frac{1}{2}y\right)} dp' dx' dy.$$

Using the Fourier inversion formula

$$\int_{\mathbb{R}^n} e^{-\frac{i}{\hbar} p' \cdot (y-x)} dp' = (2\pi\hbar)^n \delta(x-y),$$

we can rewrite $A$ as

$$A = (2\pi\hbar)^n \int_{\mathbb{R}^{2n}} \delta(x-y) e^{-\frac{i}{\hbar} p \cdot x'} \psi\left(x' + \frac{1}{2}y\right) \overline{\phi\left(x' - \frac{1}{2}y\right)} dx' dy$$

$$= (2\pi\hbar)^n \int_{\mathbb{R}^n} e^{-\frac{i}{\hbar} p \cdot x'} \psi(x' + \frac{1}{2}x) \overline{\phi\left(x' - \frac{1}{2}x\right)} dx',$$

hence $A = A(\psi, \phi)$.                                                       $\square$

**Corollary 32.** *The Fourier–Wigner transform $V(\psi, \phi)$ is the usual Fourier transform of the cross-Wigner transform:*

$$V(\psi, \phi)(p, x) = F W(\psi, \phi)(x, p). \tag{3.16}$$

**Proof.** Formula (3.16) immediately follows from (3.15).                       $\square$

The following relation shows that the cross-Wigner transform and the cross-ambiguity function essentially are the same object.

**Proposition 33.** *We have*

$$A(\psi, \phi)(z) = 2^{-n} W(\psi, \phi^\vee)\left(\frac{1}{2}z\right), \tag{3.17}$$

*where $\phi^\vee(x) = \phi(-x)$; equivalently*

$$W(\psi, \phi)(z) = 2^n A(\psi, \phi^\vee)(2z). \tag{3.18}$$

**Proof.** We have, by definition of the cross-Wigner transform,

$$W(\psi,\phi)\left(\frac{1}{2}z\right) = \left(\frac{1}{2\pi\hbar}\right)^n \int_{\mathbb{R}^n} e^{-\frac{i}{2\hbar}p\cdot y}\psi\left(\frac{1}{2}x + \frac{1}{2}y\right)\overline{\phi\left(\frac{1}{2}x - \frac{1}{2}y\right)}\,dy,$$

that is, setting $x' = \frac{1}{2}y$,

$$W(\psi,\phi^\vee)\left(\frac{1}{2}z\right) = \left(\frac{1}{\pi\hbar}\right)^n \int_{\mathbb{R}^n} e^{-\frac{i}{\hbar}p\cdot x'}\psi\left(\frac{1}{2}x + x'\right)\overline{\phi\left(\frac{1}{2}x - x'\right)}\,dx'$$

$$= \left(\frac{1}{\pi\hbar}\right)^n \int_{\mathbb{R}^n} e^{-\frac{i}{\hbar}p\cdot x'}\psi\left(x' + \frac{1}{2}x\right)\overline{\phi^\vee\left(x' - \frac{1}{2}x\right)}\,dx',$$

hence (3.17) in view of the integral expression (3.7) of $A(\psi,\phi)(z)$. $\qquad\square$

### 3.2.3. The maximum of the ambiguity function

Here is a precise boundedness result which is quite useful in radar theory (see [52, §4.2]) and which might have unexpected applications to quantum mechanics.

**Proposition 34.** *Let $\psi \in L^2(\mathbb{R}^n)$, $\psi \neq 0$. Then $z \longmapsto |A\psi(z)|$ has a unique maximum at $z = 0$:*

$$|A\psi(z)| < |A\psi(0)| = \left(\frac{1}{2\pi\hbar}\right)^n \|\psi\|_{L^2}^2$$

*for $z \neq 0$.*

**Proof.** Using the definition (3.2) of the ambiguity function, the Cauchy–Schwarz inequality yields

$$|A\psi(z_0)| = \left(\frac{1}{2\pi\hbar}\right)^n |(\psi|\widehat{T}(z_0)\psi)_{L^2}| \leq \left(\frac{1}{2\pi\hbar}\right)^n \|\psi\|_{L^2}^2$$

with equality if and only if $\widehat{T}(z_0)\psi = \lambda\psi$ for some complex constant $\lambda$; this constant has modulus one: $|\lambda| = 1$ since $\|\widehat{T}(z_0)\psi\|_{L^2}^2 = \|\psi\|_{L^2}^2$. Let us show that this can only happen when $z_0 = 0$. Suppose $z_0 \neq 0$; then $\widehat{T}(z_0)\psi = \lambda\psi$ implies, by definition (1.8) in Chapter 1 of the Heisenberg–Weyl operator, that we have

$$e^{\frac{i}{\hbar}(p_0\cdot x - \frac{1}{2}p_0\cdot x_0)}\psi(x - x_0) = \lambda\psi(x), \qquad\qquad (3.19)$$

i.e. $|\psi(x - x_0)| = |\psi(x)|$. But this implies that we must have $x_0 = 0$ (otherwise $\psi$ would be periodic, which is not possible since $\psi \in L^2(\mathbb{R}^n)$), so (3.19) reduces to

$$e^{\frac{i}{\hbar} p_0 \cdot x} \psi(x) = \lambda \psi(x).$$

But this is only possible if the factor $e^{\frac{i}{\hbar} p_0 \cdot x}$ is constant, i.e. $p_0 = 0$. $\square$

This result has the following obvious application to the Wigner transform.

**Corollary 35.** *Let* $\psi \in L^2(\mathbb{R}^n), \psi \neq 0$, *be even:* $\psi(-x) = \psi(x)$ *for all* $x \in \mathbb{R}^n$. *Then*

$$|W\psi(z)| < \left(\frac{1}{\pi\hbar}\right)^n \|\psi\|_{L^2}^2$$

*for all* $z \neq 0$.

**Proof.** It immediately follows from the proposition above noting that in view of the relation (3.18) between ambiguity function and cross-Wigner transform, we have

$$W\psi(z) = 2^n A(\psi, \psi^\vee)(2z) = 2^n A\psi(2z).$$ $\square$

## Main Formulas in Chapter 3

| | |
|---|---|
| CAF: definition 1 | $A(\psi, \phi)(z) = \left(\frac{1}{2\pi\hbar}\right)^n (\psi|\widehat{T}(z)\phi)_{L^2}$ |
| CAFT: definition 2 | $A(\psi, \phi)(z) = \left(\frac{1}{2\pi\hbar}\right)^n \int_{\mathbb{R}^n} e^{-\frac{i}{\hbar} p \cdot y} \psi(y + \frac{1}{2}x)\overline{\phi(y - \frac{1}{2}x)}dy$ |
| Ambiguity function | $A\psi(z) = \left(\frac{1}{2\pi\hbar}\right)^n \int_{\mathbb{R}^n} e^{-\frac{i}{\hbar} p \cdot y} \psi(y + \frac{1}{2}x)\overline{\psi(y - \frac{1}{2}x)}dy$ |
| Conjugation | $\overline{A(\psi, \phi)} = A(\phi^\vee, \psi^\vee)$ |
| FWT | $V(f, g)(p, x) = \int_{-\infty}^{\infty} e^{2\pi i x y} f(y + \frac{1}{2}p)\overline{g(y - \frac{1}{2}p)}dy$ |
| FWT–CAF | $V(\psi, \phi)(p, x) = A(\psi, \phi)(Jz)$ |
| CAF–CWT | $A(\psi, \phi) = F_\sigma W(\psi, \phi), \ W(\psi, \phi) = F_\sigma A(\psi, \phi)$ |
| $L^2$-boundedness | $\|A\psi\|_{L^2(\mathbb{R}^{2n})} = \left(\frac{1}{2\pi\hbar}\right)^{n/2} \|\psi\|_{L^2}$ |
| Maximum | $|A\psi(z)| < |A\psi(0)| = \left(\frac{1}{2\pi\hbar}\right)^n \|\psi\|_{L^2}^2$ if $z \neq 0$ |

# Chapter 4

# Weyl Operators

**Summary 36.** The Weyl correspondence associates to a tempered distribution ("symbol") on phase space a continuous linear operator defined on the Schwartz test functions. It can be defined using either the cross-Wigner transform or the cross-ambiguity function. In quantum mechanics, this correspondence is called the "Weyl quantization" of an observable.

In the present chapter we define the so important notion of Weyl pseudodifferential operator; we establish the fundamental relationship between the cross-Wigner transform (or the cross-ambiguity function). This remarkable and somewhat unexpected relationship justifies the fact that the part of harmonic analysis dealing with Weyl calculus is often called the Wigner–Weyl–Moyal (WWM) formalism (we used this terminology in [33]).

## 4.1. The Notion of Weyl Operator

We define in several different ways the notion of Weyl operator associated with a symbol. This correspondence between functions (or distributions) and operators is bijective, and is sometimes — somewhat unduly — called the Weyl transform.

### 4.1.1. Weyl's definition, and rigorous definitions

In what follows, the letter $a$ will generically denote an element of the distribution space $\mathcal{S}'(\mathbb{R}^{2n})$. We will call such a distribution a "symbol" (the terminology comes from the theory of pseudodifferential operators;

in time-frequency analysis one often calls $a$ the "spreading function"). Physically speaking the symbol can be viewed as an extension of the notion of "classical observable", although the latter usually is assumed to be a real function on phase space subject to some growth and regularity conditions.

Historically the formal definition of the Weyl operator with symbol $a$ goes as follows (see [80]): in view of the Fourier inversion formula, we can write

$$a(x,p) = \left(\frac{1}{2\pi\hbar}\right)^n \int_{\mathbb{R}^{2n}} Fa(x_0, p_0) e^{\frac{i}{\hbar}(x_0 \cdot x + p_0 \cdot p)} dx_0 dp_0.$$

One now wants to replace the variables $x$ and $p$ by some vector-valued operators $\hat{x} = (\hat{x}_1, \ldots, \hat{x}_n)$ and $\hat{p} = (\hat{p}_1, \ldots, \hat{p}_n)$ satisfying the canonical commutation relations $[\hat{x}_j, \hat{p}_k] = i\hbar\delta_{jk}$; a standard choice in quantum mechanics is to take for $\hat{x}_j$ the operator of multiplication by the coordinate $x_j$ and $\hat{p}_j = -i\hbar\partial_{x_j}$ (a theorem of von Neumann anyway ensures us that all other possible choices are unitarily equivalent to this one, so we do not have to bother too much; other choices are however possible — and useful! — as we will see in Chapter 11 when we study the Schrödinger equation in phase space). Denoting by $\hat{A}$ the operator $a(\hat{x}, \hat{p})$ thus formally obtained we have

$$\hat{A} = \left(\frac{1}{2\pi\hbar}\right)^n \int_{\mathbb{R}^{2n}} Fa(x_0, p_0) e^{\frac{i}{\hbar}(x_0 \cdot \hat{x} + p_0 \cdot \hat{x})} dx_0 dp_0.$$

The problem is of course that we have to give a precise meaning to the exponential $e^{\frac{i}{\hbar}(x_0 \cdot \hat{x} + p_0 \cdot \hat{x})}$ in expressions of the type above. But this we have already done in Chapter 1 where we defined

$$\widehat{M}(z_0) = e^{\frac{i}{\hbar}(x_0 \cdot \hat{x} + p_0 \cdot \hat{x})}$$

by the formula

$$\widehat{M}(z_0)\psi(x) = e^{\frac{i}{\hbar}(x_0 \cdot x + \frac{1}{2}p_0 \cdot x_0)}\psi(x + p_0).$$

We also showed that the Heisenberg–Weyl operator and the exponential $\widehat{M}(z_0)$ are related by $\widehat{T}(-Jz_0) = \widehat{M}(z_0)$. This suggests the following correct mathematical definition.

**Definition 37.** Let $a \in \mathcal{S}(\mathbb{R}^{2n})$. The Weyl operator $\hat{A} = \mathrm{Op}_W(a)$ with symbol $a$ is defined, for $\psi \in \mathcal{S}(\mathbb{R}^n)$, by

$$\hat{A}\psi(x) = \left(\frac{1}{2\pi\hbar}\right)^n \int_{\mathbb{R}^{2n}} Fa(z_0)\widehat{M}(z_0)\psi(x)dz_0 \qquad (4.1)$$

or, equivalently, by

$$\widehat{A}\psi(x) = \left(\frac{1}{2\pi\hbar}\right)^n \int_{\mathbb{R}^{2n}} a_\sigma(z_0)\widehat{T}(z_0)\psi(x)dz_0, \tag{4.2}$$

where $a_\sigma = F_\sigma a$ is the symplectic Fourier transform of $a$.

Both definitions are indeed equivalent: making the change of variables $z_0 \longmapsto Jz_0$ in (4.1) yields

$$\widehat{A}\psi(x) = \left(\frac{1}{2\pi\hbar}\right)^n \int_{\mathbb{R}^{2n}} Fa(Jz_0)\widehat{M}(Jz_0)\psi(x)dz_0,$$

which is (4.2) since $\widehat{T}(z_0) = \widehat{M}(Jz_0)$ and $Fa(Jz_0) = F_\sigma a(z_0)$.

**Remark 38.** Notice that we have $\widehat{A} = 0$ if and only if $a = 0$: if $\widehat{A}\psi = 0$ for all $\psi \in \mathcal{S}(\mathbb{R}^n)$ then $Fa = 0$ and hence $a = 0$ (and conversely). The Weyl correspondence is thus injective, as far as symbols in $\mathcal{S}(\mathbb{R}^{2n})$ are considered (we will see that it is so as well in the general case).

We will often write the definitions above without referring to the function on which $\widehat{A}$ acts:

$$\widehat{A} = \left(\frac{1}{2\pi\hbar}\right)^n \int_{\mathbb{R}^{2n}} Fa(z_0)\widehat{M}(z_0)dz_0 \tag{4.3}$$

or, equivalently,

$$\widehat{A} = \left(\frac{1}{2\pi\hbar}\right)^n \int_{\mathbb{R}^{2n}} a_\sigma(z_0)\widehat{T}(z_0)dz_0 \tag{4.4}$$

(these integrals can be viewed as "Bochner integrals", i.e. vector-valued integrals). The symplectic Fourier transform $F_\sigma a$ of the symbol is an important object, appearing often in both quantum mechanics and in time-frequency analysis.

**Definition 39.** The symplectic Fourier transform $a_\sigma = F_\sigma a$ of the Weyl symbol of the operator $\widehat{A}$ is called the twisted symbol of $\widehat{A}$ (or sometimes its "covariant symbol").

Both definitions (4.3) and (4.4) express the Weyl correspondence in terms of the Fourier transform of the symbol. Using the Grossmann–Royer operator, we can also express $\widehat{A}$ in terms of the symbol $a$ itself.

**Proposition 40.** *Let* $a \in \mathcal{S}(\mathbb{R}^{2n})$. *The Weyl operator* $\widehat{A} = \mathrm{Op_W}(a)$ *is given for* $\psi \in \mathcal{S}(\mathbb{R}^n)$ *by*

$$\widehat{A}\psi(x) = \left(\frac{1}{\pi\hbar}\right)^n \int_{\mathbb{R}^{2n}} a(z_0)\widehat{R}(z_0)\psi(x)dz_0, \qquad (4.5)$$

*where* $\widehat{R}(z_0)$ *is the Grossmann–Royer operator.*

**Proof.** Recall from Proposition 8 that the Grossmann–Royer operator is related to the Heisenberg–Weyl operator by the formula

$$\widehat{R}(z_0)\psi(x) = 2^{-n}F_\sigma(\widehat{T}(\cdot)\psi(x))(-z_0),$$

where $x$ is fixed. Since the symplectic Fourier transform is an involution, this equality can be rewritten as

$$\widehat{T}(z_0)\psi(x) = 2^n F_\sigma(\widehat{R}(\cdot)\psi(x))(-z_0).$$

Insertion in (4.2) yields

$$\widehat{A}\psi(x) = \left(\frac{1}{\pi\hbar}\right)^n \int_{\mathbb{R}^{2n}} a_\sigma(z_0)F_\sigma(\widehat{R}(\cdot)\psi(x))(-z_0))\psi(x)dz_0.$$

Using the Plancherel formula (see Appendix B), we can rewrite this as follows:

$$\widehat{A}\psi(x) = \left(\frac{1}{\pi\hbar}\right)^n \int_{\mathbb{R}^{2n}} F_\sigma a_\sigma(z_0)F_\sigma^2(\widehat{R}(\cdot)\psi(x))(z_0))\psi(x)dz_0,$$

that is, since $F_\sigma$ is an involution,

$$\widehat{A}\psi(x) = \left(\frac{1}{\pi\hbar}\right)^n \int_{\mathbb{R}^{2n}} a(z_0)\widehat{R}(z_0)\psi(x)dz_0$$

proving formula (4.5). $\qquad\qquad\square$

Weyl operators are continuous on their domain, and can be extended to symbols which are tempered distributions. we introduce the following notation.

**Notation 41.** $\mathcal{L}(\mathcal{S}(\mathbb{R}^n), \mathcal{S}'(\mathbb{R}^n))$ is the space of continuous linear mappings $\mathcal{S}(\mathbb{R}^n) \longrightarrow \mathcal{S}'(\mathbb{R}^n)$.

**Proposition 42.** *Let* $a \in \mathcal{S}(\mathbb{R}^{2n})$. *Then we have the following statements.*

(i) *The Weyl operator* $\widehat{A} = \mathrm{Op_W}(a)$ *is a continuous operator* $\mathcal{S}(\mathbb{R}^n) \longrightarrow \mathcal{S}(\mathbb{R}^n)$.

(ii) *The Weyl correspondence* $a \longmapsto \mathrm{Op_W}(a)$ *extends into a continuous mapping*

$$\mathrm{Op_W} : \mathcal{S}'(\mathbb{R}^{2n}) \longrightarrow \mathcal{L}(\mathcal{S}(\mathbb{R}^n), \mathcal{S}'(\mathbb{R}^n)).$$

**Proof.** Writing formula (4.5) using distributional brackets we have

$$\widehat{A}\psi = \left(\frac{1}{\pi\hbar}\right)^n \langle a(\cdot), \widehat{R}(\cdot)\psi \rangle.$$

The statements follows. $\qquad\qquad\qquad\qquad\qquad\qquad\qquad\qquad\qquad\square$

We leave it to the reader to check the following elementary formulas:

$$\mathrm{Op_W}(x_j)\psi = x_j\psi, \quad \mathrm{Op_W}(p_j)\psi = -i\hbar\partial_{x_j}\psi.$$

More generally, we have the following monomial quantization rules.

**Example 43.** Let $r$ and $s$ be non-negative integers. We have

$$\mathrm{Op_W}(x_j^r p_j^s) = \frac{1}{2^s} \sum_{\ell=0}^{s} \binom{s}{\ell} \widehat{p}_j^{\,s-\ell} \widehat{x}_j^{\,r} \widehat{p}_j^{\,\ell}; \qquad (4.6)$$

equivalently

$$\mathrm{Op_W}(x_j^r p_j^s) = \frac{1}{2^s} \sum_{\ell=0}^{s} \binom{s}{\ell} \widehat{x}_j^{\,s-\ell} \widehat{p}_j^{\,r} \widehat{x}_j^{\,\ell}; \qquad (4.7)$$

this relation can be rewritten in several different ways, for instance

$$\mathrm{Op_W}(x_j^r p_j^s) = \sum_{\ell=0}^{\min(r,s)} (-i\hbar)^\ell \binom{s}{\ell} \binom{r}{\ell} \frac{\ell!}{2^\ell} \widehat{x}_j^{\,r-\ell} \widehat{p}_j^{\,s-\ell}; \qquad (4.8)$$

$$\mathrm{Op_W}(x^r p^s) = \sum_{\ell=0}^{\min(r,s)} (i\hbar)^\ell \binom{s}{\ell} \binom{r}{\ell} \frac{\ell!}{2^\ell} \widehat{p}_j^{\,s-\ell} \widehat{x}_j^{\,r-\ell}. \qquad (4.9)$$

(See [41].)

## 4.1.2. The distributional kernel of a Weyl operator

The formula (4.5) allows us to write the definition of $\widehat{A}$ in the usual textbook integral form. In fact, using the explicit definition (1.15) in Chapter 1 of the Grossmann–Royer operator, we have

$$\widehat{R}(z_0)\psi(x) = e^{\frac{2i}{\hbar}p_0\cdot(x-x_0)}\psi(2x_0 - x),$$

and hence

$$\widehat{A}\psi(x) = \left(\frac{1}{\pi\hbar}\right)^n \int_{\mathbb{R}^{2n}} a(z_0)e^{\frac{2i}{\hbar}p_0\cdot(x-x_0)}\psi(2x_0 - x)dz_0.$$

Setting $y = 2x_0 - x$ we have $x - x_0 = x_0 - y$ that is, replacing $p_0$ with $p$,

$$\widehat{A}\psi(x) = \left(\frac{1}{2\pi\hbar}\right)^n \int_{\mathbb{R}^{2n}} e^{\frac{i}{\hbar}p\cdot(x-y)}a\left(\frac{1}{2}(x + y), p\right)\psi(y)dydp \qquad (4.10)$$

which is the usual integral expression often used as the definition of a Weyl operator in the theory of pseudodifferential operators [28, 76].

While formula (4.10) strictly speaking only makes sense under rather stringent assumptions on the symbol $a$ and the function $\psi$ (because of obvious problems of convergence of the integral), it can be given a rigorous meaning in the general case in the distributional sense. Let us first recall Schwartz's kernel theorem (see, e.g. [52] for a modern presentation): it says that for every continuous operator $\widehat{A}: \mathcal{S}(\mathbb{R}^n) \longrightarrow \mathcal{S}'(\mathbb{R}^n)$ there exists a distribution $K \in \mathcal{S}'(\mathbb{R}^n \times \mathbb{R}^n)$ (the distributional kernel of $\widehat{A}$) such that

$$\langle\widehat{A}\psi, \phi\rangle = \langle K, \phi \otimes \psi\rangle \qquad (4.11)$$

for all $\psi, \phi \in \mathcal{S}(\mathbb{R}^n)$. Committing a minor abuse of notation this is often written explicitly in integral notation as

$$\widehat{A}\psi(x) = \int_{\mathbb{R}^n} K(x, y)\psi(y)dy.$$

Comparing with the expression (4.10) for $\widehat{A}\psi(x)$ we see that the distributional kernel of $\widehat{A}$ is formally given by

$$K(x, y) = \left(\frac{1}{2\pi\hbar}\right)^n \int_{\mathbb{R}^n} e^{\frac{i}{\hbar}p\cdot(x-y)}a\left(\frac{1}{2}(x + y), p\right)dp; \qquad (4.12)$$

this formula can in fact be made rigorous if we *define* $\widehat{A}$ as being the operator $\mathcal{S}(\mathbb{R}^n) \longrightarrow \mathcal{S}'(\mathbb{R}^n)$ with kernel

$$K(x, y) = \left(\frac{1}{2\pi\hbar}\right)^{n/2} F_2^{-1}a\left(\frac{1}{2}(x + y), x - y\right), \qquad (4.13)$$

where $F_2^{-1}$ is the inverse Fourier transform in the set of variables $p$.

**Remark 44.** This formula allows us also to define the Weyl operator for $a \in \mathcal{S}'(\mathbb{R}^{2n})$ as the operator whose kernel is given by the (inverse) partial Fourier transform (4.12).

Notice that the Fourier inversion formula then allows us conversely to express the symbol $a$ in terms of the kernel $K$:

$$a(x, p) = \int_{\mathbb{R}^n} e^{-\frac{i}{\hbar} p \cdot y} K\left(x + \frac{1}{2}y, x - \frac{1}{2}y\right) dy. \tag{4.14}$$

This formula should be understood as a partial Fourier transform: let $V$ be the linear diffeomorphism $(x, y) \longmapsto (x + \frac{1}{2}y, x - \frac{1}{2}y)$; then

$$K\left(x + \frac{1}{2}y, x - \frac{1}{2}y\right) = K \circ V(x, y)$$

so that

$$a(x, p) = \int_{\mathbb{R}^n} e^{-\frac{i}{\hbar} p \cdot y} K \circ F(x, y) dy = (2\pi\hbar)^{n/2} F_2(K \circ V).$$

Summarizing the discussion above we see that this correspondence between Weyl symbol and distributional kernel implies that every continuous operator is a Weyl operator.

**Proposition 45.** *For every continuous operator* $\widehat{A} : \mathcal{S}(\mathbb{R}^n) \longrightarrow \mathcal{S}'(\mathbb{R}^n)$, *there exists a unique* $a \in \mathcal{S}'(\mathbb{R}^{2n})$ *such that* $\widehat{A} = \mathrm{Op_W}(a)$.

**Proof.** In view of Schwartz's kernel theorem every continuous operator $\widehat{A} : \mathcal{S}(\mathbb{R}^n) \longrightarrow \mathcal{S}'(\mathbb{R}^n)$ has a distributional kernel $K \in \mathcal{S}'(\mathbb{R}^n \times \mathbb{R}^n)$. Defining the symbol $a$ by formula (4.14) we have $\widehat{A} = \mathrm{Op_W}(a)$. $\square$

**Example 46.** The Weyl symbol of the operator with kernel $K = (2\pi\hbar)^{-n} \psi \otimes \overline{\psi}$ is the Wigner transform $W\psi$. More generally, $K = (2\pi\hbar)^{-n} \psi \otimes \overline{\phi}$ the corresponding Weyl symbol is the cross-Wigner transform $W(\psi, \phi)$. This is an immediate consequence of formula (4.14) and the integral definition of $W\psi$.

The example above suggests there is a close relationship between Weyl operators and the cross-Wigner transform. This is indeed the case that is shown in next section. Before establishing this correspondence, let us prove a result which will be useful when we study the mixed states of quantum mechanics in Chapter 13.

**Proposition 47.** *Let* $\widehat{A} = \mathrm{Op_W}(a)$ *have distributional kernel* $K \in L^2(\mathbb{R}^n \times \mathbb{R}^n)$. *We then have* $a \in L^2(\mathbb{R}^{2n})$ *and*

$$||a||_{L^2(\mathbb{R}^{2n})} = (2\pi\hbar)^{n/2} ||K||_{L^2(\mathbb{R}^n \times \mathbb{R}^n)}. \qquad (4.15)$$

**Proof.** (Cf. [39, Proposition 209]). Let us first prove that $||a||_{L^2} = (2\pi\hbar)^{n/2} ||K_{\widehat{A}}||_{L^2}$ when $K \in \mathcal{S}(\mathbb{R}^n \times \mathbb{R}^n)$. In view of formula (4.13) the symbol $a$ is, for fixed $x$, $(2\pi\hbar)^{n/2}$ times the Fourier transform of the function $y \longmapsto K(x + \frac{1}{2}y, x - \frac{1}{2}y)$ hence, by Plancherel's formula,

$$\int_{\mathbb{R}^n} |a(x,p)|^2 dp = (2\pi\hbar)^n \int_{\mathbb{R}^n} \left| K\left(x + \frac{1}{2}y, x - \frac{1}{2}y\right)\right|^2 dy \qquad (4.16)$$

and hence

$$\int_{\mathbb{R}^{2n}} |a(z)|^2 dz = (2\pi\hbar)^n \int_{\mathbb{R}^n} \left( \int_{\mathbb{R}^n} \left| K\left(x + \frac{1}{2}y, x - \frac{1}{2}y\right)\right|^2 dy \right) dx$$

$$= (2\pi\hbar)^n \int_{\mathbb{R}^{2n}} \left| K\left(x + \frac{1}{2}y, x - \frac{1}{2}y\right)\right|^2 dxdy,$$

where we have applied Fubini's theorem (the integrals are absolutely convergent since the function

$$(x,y) \longmapsto K\left(x + \frac{1}{2}y, x - \frac{1}{2}y\right)$$

is in $\mathcal{S}(\mathbb{R}^n \times \mathbb{R}^n)$ because the kernel $K$ is). Set now $x' = x + \frac{1}{2}y$ and $y' = x - \frac{1}{2}y$; we have $dx'dy' = dxdy$ hence

$$\int_{\mathbb{R}^{2n}} |a(z)|^2 dz = (2\pi\hbar)^n \int_{\mathbb{R}^{2n}} |K_{\widehat{A}}(x', y')|^2 dx'dy'$$

proving (4.15) for $K \in \mathcal{S}(\mathbb{R}^n \times \mathbb{R}^n)$; the general case follows since $\mathcal{S}(\mathbb{R}^n \times \mathbb{R}^n)$ is dense in $L^2(\mathbb{R}^n \times \mathbb{R}^n)$. $\qquad \square$

**Remark 48.** The result above actually characterizes the symbols of Hilbert–Schmidt operators on the Hilbert space $L^2(\mathbb{R}^n)$: the latter are exactly those operators having a square-integrable kernel (or a square-integrable Weyl symbol).

### 4.1.3. Relation with the cross-Wigner transform

As far we have defined Weyl operators for a very limited class of symbols, those belonging to the Schwartz class $\mathcal{S}(\mathbb{R}^{2n})$. It turns out that this definition can be extended to all symbols $a \in \mathcal{S}'(\mathbb{R}^{2n})$. This can be shown in several ways; the method we will use here is based on the fundamental relation between the Weyl correspondence and the cross-Wigner transform.

**Proposition 49.** *Let* $a \in \mathcal{S}(\mathbb{R}^{2n})$. *We have*

$$(\widehat{A}\psi|\phi)_{L^2} = \int_{\mathbb{R}^{2n}} a(z)W(\psi,\phi)(z)dz \tag{4.17}$$

*for all* $\psi, \phi \in \mathcal{S}(\mathbb{R}^n)$; *more generally, if* $a \in \mathcal{S}'(\mathbb{R}^{2n})$, *then*

$$\langle \widehat{A}\psi, \overline{\phi} \rangle = \langle a, W(\psi,\phi) \rangle. \tag{4.18}$$

**Proof.** Using formula (4.10) we have

$$(\widehat{A}\psi|\phi)_{L^2} = \left(\frac{1}{\pi\hbar}\right)^n \int_{\mathbb{R}^{3n}} e^{\frac{i}{\hbar}p\cdot(x-y)} a\left(\frac{1}{2}(x+y),p\right)\psi(y)\overline{\phi(x)}dydpdx;$$

performing the unitary change of variables $(x,y) \longmapsto (x + \frac{1}{2}y, x - \frac{1}{2}y)$ (cf. the operator $V$ defined by (2.16) in Proposition 16), this equality becomes

$$(\widehat{A}\psi|\phi)_{L^2} = \left(\frac{1}{\pi\hbar}\right)^n \int_{\mathbb{R}^{3n}} e^{-\frac{i}{\hbar}p\cdot y} a(x,p)\psi\left(x + \frac{1}{2}y\right)\overline{\phi\left(x - \frac{1}{2}y\right)}dydpdx;$$

the formula (4.17) follows by Fubini's theorem. Formula (4.18) follows since $W(\psi,\phi) \in \mathcal{S}(\mathbb{R}^{2n})$ if $\psi, \phi \in \mathcal{S}(\mathbb{R}^n)$. □

**Remark 50.** Formula (4.17) can be taken — and often is — as the definition of the Weyl operator $\widehat{A} = \text{Op}_W(a)$.

As an application let us find the symbol of a Heisenberg–Weyl operator.

**Proposition 51.** *We have* $\widehat{T}(z_0) = \text{Op}_W(e^{-\frac{i}{\hbar}\sigma(\cdot,z_0)})$. *In particular,* $\text{Op}_W(1) = I_d$ *(the identity operator).*

**Proof.** The statement $\mathrm{Op_W}(1) = I_\mathrm{d}$ is a particular case of the relation

$$\widehat{T}(z_0) = \mathrm{Op_W}(e^{-\frac{i}{\hbar}\sigma(\cdot,z_0)})$$

taking $z_0 = 0$. Let us write $a(z) = e^{-\frac{i}{\hbar}\sigma(z,z_0)}$ and set $\widehat{A} = \mathrm{Op_W}(a)$. We have, for $\psi \in \mathcal{S}(\mathbb{R}^n)$,

$$\widehat{A}\psi(x) = \left(\frac{1}{\pi\hbar}\right)^n \int_{\mathbb{R}^{2n}} e^{-\frac{i}{\hbar}\sigma(z,z')} e^{\frac{2i}{\hbar}p_0\cdot(x-x_0)} \psi(2x_0 - x) dp_0 dx_0$$

$$= \left(\frac{1}{\pi\hbar}\right)^n \int_{\mathbb{R}^{2n}} e^{\frac{i}{\hbar}p'\cdot x_0} e^{\frac{i}{\hbar}p_0\cdot(x'+2x_0-2x)} \psi(2x_0 - x) dp_0 dx_0$$

$$= \left(\frac{1}{\pi\hbar}\right)^n \int_{\mathbb{R}^n} e^{\frac{i}{\hbar}p'\cdot x_0} \left(\int_{\mathbb{R}^n} e^{\frac{i}{\hbar}p_0\cdot(x'+2x_0-2x)} dp_0\right) \psi(2x_0 - x) dx_0$$

$$= 2^n \int_{\mathbb{R}^n} e^{\frac{i}{\hbar}p'\cdot x_0} \delta(x' + 2x_0 - 2x)\psi(2x_0 - x) dx_0.$$

Setting $y = 2x_0$ we get

$$\widehat{A}\psi(x) = \int_{\mathbb{R}^n} e^{\frac{i}{\hbar}p'\cdot x_0} \delta(y - (2x - x'))\psi(2x_0 - x) dx_0$$

$$= e^{\frac{i}{\hbar}(p'\cdot x - \frac{1}{2}p'\cdot x')}\psi(x - x')$$

which concludes the proof of the proposition.                               $\square$

Not surprisingly, the Wigner transform is itself the Weyl symbol of an operator.

**Proposition 52.** *Let $\psi \in L^2(\mathbb{R}^n)$, $\psi \neq 0$. The Weyl symbol of the orthogonal projection $\widehat{\Pi}_\psi$ on the ray $\{\lambda\psi : \lambda \in \mathbb{C}\}$ in $L^2(\mathbb{R}^n)$ generated by $\psi$ is given by the formula*

$$\pi_\psi(z) = (2\pi\hbar)^n W\psi(z). \tag{4.19}$$

**Proof.** We have, by definition of the projector $\Pi_\psi$,

$$\Pi_\psi \phi(x) = \int_{\mathbb{R}^n} \psi(x)\overline{\phi(y)}\psi(y) dy,$$

and hence the kernel of $\widehat{\Pi}_\psi$ is $K = \psi \otimes \overline{\phi}$. In view of the relation (4.14) between kernel and symbol, the symbol $\pi_\psi$ of $\widehat{\Pi}_\psi$ is thus given by

$$\pi_\psi(x, p) = \int_{\mathbb{R}^n} e^{-\frac{i}{\hbar} p \cdot y} \psi\left(x + \frac{1}{2} y\right) \overline{\phi\left(x - \frac{1}{2} y\right)} dy$$

hence our claim. $\qquad\square$

## 4.2. Some Properties of the Weyl Correspondence

It is out of question to give a complete account of the regularity and analytical properties of Weyl operators in a single chapter. There exists a rich literature about the topic; here are a few suggestions (in alphabetical order) from my own library: Folland [28], Gröchenig [52], Wong [82], Stein [78]. We have also given shorter accounts in our previous monographs [33, 39].

We limit ourselves here to two topics, each of them important in their own right: the adjoint of a Weyl operator and the $L^2$ boundedness of Weyl operators with symbols whose Fourier transforms are absolutely integrable. The property of symplectic covariance of the Weyl correspondence, which is a characteristic property of Weyl calculus, will be addressed separately in Chapter 5.

### 4.2.1. The adjoint of a Weyl operator

Let $\widehat{A} : \mathcal{S}(\mathbb{R}^n) \longrightarrow \mathcal{S}(\mathbb{R}^n)$ be a continuous operator. The formal adjoint $\widehat{A}^*$ with respect to the sesquilinear product $(\cdot|\cdot)_{L^2}$ is defined by

$$(\widehat{A}\psi|\phi)_{L^2} = (\psi|\widehat{A}^*\phi)_{L^2}$$

for all $\psi, \phi \in \mathcal{S}(\mathbb{R}^n), \mathcal{S}(\mathbb{R}^n)$. We are going to determine explicitly the Weyl correspondence $\widehat{A}^* \overset{\text{Weyl}}{\longleftrightarrow} a^*$. For this we need the following elementary lemma.

**Lemma 53.** *Let $b$ be a function on $\mathbb{R}^n$ such that $(b|W(\psi, \phi))_{L^2} = 0$ for all $\psi, \phi \in \mathcal{S}(\mathbb{R}^n)$. Then $b = 0$.*

**Proof.** In view of formula (4.17) we have, noting that $\overline{W(\psi, \phi)} = W(\phi, \psi)$,

$$(\widehat{B}\phi|\psi)_{L^2} = (b|W(\psi, \phi))_{L^2(\mathbb{R}^{2n})},$$

where $\widehat{B} = \mathrm{Op_W}(b)$. If $(b|W(\psi, \phi))_{L^2(\mathbb{R}^{2n})} = 0$ for all $\psi$ and $\phi$, we must thus have $\widehat{B}\phi = 0$ for all $\phi$ hence $\widehat{B} = 0$; but then $b = 0$ since the Weyl correspondence is bijective.                                                                                     $\square$

**Proposition 54.** *The formal adjoint* $\widehat{A}^*$ *of* $\widehat{A} = \mathrm{Op_W}(a)$ *is the Weyl operator* $\widehat{A}^* = \mathrm{Op_W}(\overline{a})$. *In particular,* $\widehat{A}$ *is (formally) self-adjoint if and only if the symbol* $a$ *is real.*

**Proof.** Using formula we have

$$(\widehat{A}\psi|\phi)_{L^2} = \left(\frac{1}{\pi\hbar}\right)^n \int_{\mathbb{R}^{2n}} a(z_0) \left(\int_{\mathbb{R}^n} \widehat{R}(z_0)\psi(x)\overline{\phi(x)}\right) dz_0$$

$$= \int_{\mathbb{R}^{2n}} a(z_0) W(\psi, \phi)(z_0) dz_0.$$

Let us denote by $a^*$ the Weyl symbol of $\widehat{A}^*$, that is $\widehat{A}^* = \mathrm{Op_W}(a^*)$ (this symbol exists in view of our discussion above). We have $(\psi|\widehat{A}^*\phi)_{L^2} = \overline{(\widehat{A}^*\phi|\psi)_{L^2}}$ and hence, using formula (4.17),

$$(\psi|\widehat{A}^*\phi)_{L^2} = \int_{\mathbb{R}^{2n}} \overline{a^*}(z_0)\overline{W(\phi, \psi)}(z_0) dz_0$$

$$= \int_{\mathbb{R}^{2n}} \overline{a^*}(z_0) W(\psi, \phi)(z_0) dz_0.$$

Since by definition $(\psi|\widehat{A}^*\phi)_{L^2} = (\widehat{A}\psi|\phi)_{L^2}$ we thus have

$$(\widehat{A}\psi|\phi)_{L^2} = \int_{\mathbb{R}^{2n}} \overline{a^*}(z_0) W(\psi, \phi)(z_0) dz_0$$

for all $\psi, \phi \in \mathcal{S}(\mathbb{R}^n)$ hence $a^* = \overline{a}$ applying Lemma 53 to $b = a - \overline{a^*}$.     $\square$

### 4.2.2. An $L^2$ boundedness result

The following boundedness result is interesting when on studies the density operator of a mixed quantum state.

**Proposition 55.** *Let* $a \in \mathcal{S}'(\mathbb{R}^{2n})$ *and assume that* $Fa \in L^1(\mathbb{R}^{2n})$ *(F the Fourier transform on* $\mathbb{R}^{2n}$*). Then the operator* $\widehat{A} = \mathrm{Op_W}(a)$ *is bounded on* $L^2(\mathbb{R}^n)$ *and we have*

$$||\widehat{A}\psi||_{L^2} \leq \left(\frac{1}{2\pi\hbar}\right)^{2n} ||Fa||_{L^1(\mathbb{R}^{2n})} ||\psi||_{L^2}. \tag{4.20}$$

**Proof.** It suffices to prove that the estimate (4.20) holds for all $\psi \in \mathcal{S}(\mathbb{R}^n)$; the general case $\psi \in L^2(\mathbb{R}^n)$ follows since $\mathcal{S}(\mathbb{R}^n)$ is dense in $L^2(\mathbb{R}^n)$. Recall (formula (4.13)) that the kernel of $\widehat{A}$ is given by

$$K(x,y) = \left(\frac{1}{2\pi\hbar}\right)^{n/2} F_2^{-1}a\left(\frac{1}{2}(x+y), x-y\right)$$

where $F_2$ is the Fourier transform in the $p$ variables. By the Fourier inversion formula, we have

$$F_2^{-1}a\left(\frac{1}{2}(x+y), y-x\right) = \left(\frac{1}{2\pi\hbar}\right)^{n/2} \int_{\mathbb{R}^n} e^{-\frac{i}{2\hbar}(x+y)\cdot\xi}F^{-1}a(\xi, x-y)d\xi$$

and hence

$$K(x,y) = \left(\frac{1}{2\pi\hbar}\right)^{n} \int_{\mathbb{R}^n} e^{-\frac{i}{2\hbar}(x+y)\cdot\xi}Fa(\xi, x-y)d\xi.$$

It follows that

$$\int |K(x,y)|dx \leq \left(\frac{1}{2\pi\hbar}\right)^{n} \int_{\mathbb{R}^{2n}} |Fa(\xi, x-y)|d\xi dx,$$

$$\int |K(x,y)|dy \leq \left(\frac{1}{2\pi\hbar}\right)^{n} \int_{\mathbb{R}^{2n}} |Fa(\xi, x-y)|d\xi dy.$$

Setting $\eta = y - x$, we have

$$\int_{\mathbb{R}^{2n}} |Fa(\xi, y-x)|d\xi dx = \int_{\mathbb{R}^{2n}} |Fa(\xi, \eta)|d\xi d\eta,$$

$$\int_{\mathbb{R}^{2n}} |Fa(\xi, y-x)|d\xi dy = \int_{\mathbb{R}^{2n}} |Fa(\xi, \eta)|d\xi d\eta,$$

and hence the two inequalities above can be rewritten as

$$\int_{\mathbb{R}^n} |K(x,y)|dx \leq \left(\frac{1}{2\pi\hbar}\right)^{n} ||Fa||_{L^1}, \qquad (4.21)$$

$$\int_{\mathbb{R}^n} |K(x,y)|dy \leq \left(\frac{1}{2\pi\hbar}\right)^{n} ||Fa||_{L^1}. \qquad (4.22)$$

Expressing the operator $\widehat{A}$ in terms of the kernel we have

$$\widehat{A}\psi(x) = \int_{\mathbb{R}^n} K(x,y)\psi(y)dy$$

hence

$$|\widehat{A}\psi(x)| \le \int_{\mathbb{R}^n} |K(x,y)|\,|\psi(y)|dx.$$

Writing $|K| = |K|^{1/2}|K|^{1/2}$, it follows from Cauchy–Schwarz's inequality and (4.22) that

$$|\widehat{A}\psi(x)|^2 \le \int_{\mathbb{R}^n} |K(x,y)|dy \int_{\mathbb{R}^n} |K(x,y)||\psi(y)|^2 dy$$

$$\le \left(\frac{1}{2\pi\hbar}\right)^n ||Fa||_{L^1} \int_{\mathbb{R}^n} |K(x,y)||\psi(y)|^2 dy,$$

and hence

$$\int_{\mathbb{R}^n} |\widehat{A}\psi(x)|^2 dx \le \left(\frac{1}{2\pi\hbar}\right)^n ||Fa||_{L^1} \int_{\mathbb{R}^n} \left(\int_{\mathbb{R}^n} |K(x,y)|dx\right) |\psi(y)|^2 dy.$$

Using the inequality (4.21), we get

$$\int_{\mathbb{R}^n} \left(\int_{\mathbb{R}^n} |K(x,y)|dx\right) |\psi(y)|^2 dy \le \left(\frac{1}{2\pi\hbar}\right)^n ||Fa||_{L^1} \int_{\mathbb{R}^n} |\psi(y)|^2 dy;$$

summarizing

$$\int_{\mathbb{R}^n} |\widehat{A}\psi(x)|^2 dx \le \left(\frac{1}{2\pi\hbar}\right)^{2n} ||Fa||_{L^1}^2 \int_{\mathbb{R}^n} |\psi(y)|^2 dy$$

which is precisely the estimate (4.20). $\qquad\qquad\square$

## Main Formulas in Chapter 4

| | |
|---|---|
| Weyl Operator 1 | $\widehat{A}\psi(x) = \left(\frac{1}{2\pi\hbar}\right)^n \int_{\mathbb{R}^{2n}} Fa(z_0)e^{\frac{i}{\hbar}\widehat{z}\cdot z_0}\psi(x)dz_0$ |
| Weyl Operator 2 | $\widehat{A}\psi(x) = \left(\frac{1}{2\pi\hbar}\right)^n \int_{\mathbb{R}^{2n}} a_\sigma(z_0)\widehat{T}(z_0)\psi(x)dz_0$ |
| Weyl Operator 3 | $\widehat{A}\psi(x) = \left(\frac{1}{\pi\hbar}\right)^n \int_{\mathbb{R}^{2n}} a(z_0)\widehat{R}(z_0)\psi(x)dz_0$ |
| Monomials | $\widehat{x_j^r p_j^s} = \frac{1}{2^s}\sum_{\ell=0}^s \binom{s}{\ell}\widehat{p}_j^{\,s-\ell}\widehat{x}_j^{\,r}\widehat{p}_j^{\,\ell}$ |
| Kernel/Symbol | $K(x,y) = \left(\frac{1}{2\pi\hbar}\right)^n \int_{\mathbb{R}^n} e^{\frac{i}{\hbar}p\cdot(x-y)}a(\tfrac{1}{2}(x+y),p)dp$ |
| Symbol/Kernel | $a(x,p) = \int_{\mathbb{R}^n} e^{-\frac{i}{\hbar}p\cdot y}K(x+\tfrac{1}{2}y, x-\tfrac{1}{2}y)dy$ |
| Weyl and CWT | $(\widehat{A}\psi|\phi)_{L^2} = \int_{\mathbb{R}^{2n}} a(z)W(\psi,\phi)(z)dz$ |
| Adjoint | $\widehat{A} = \mathrm{Op_W}(a) \implies \widehat{A}^* = \mathrm{Op_W}(\overline{a})$ |
| Symplectic covariance | $\mathrm{Op_W}(a \circ S) = \widehat{S}^{-1}\mathrm{Op_W}(a)\widehat{S}$ |

# Chapter 5

# Symplectic Covariance

**Summary 56.** The metaplectic group is a group of unitary operators. Symplectic changes of coordinates in the cross-Wigner transform and in the cross-ambiguity function correspond to conjugation by metaplectic operators. These results are optimal in the following sense: only symplectic and antisymplectic transformations ensure such a linear covariance.

In any mathematical theory, symmetries play an essential role, not only because they allow to simplify many problems, but also because they reveal the structure of that theory. The symplectic group is such a symmetry group in the Weyl–Wigner–Moyal theory.

## 5.1. Symplectic Covariance Properties

We shortly collect the symplectic material we will need in this chapter. For details, we refer the reader to Appendix A and to the references therein.

### 5.1.1. Review of some properties of Mp(n) and Sp(n)

Recall that the symplectic group $\mathrm{Sp}(n)$ consists of all linear automorphisms of phase space $\mathbb{R}^{2n} = \mathbb{R}^n_x \times \mathbb{R}^n_p$ which preserve the symplectic form $\sigma(z, z') = Jz \cdot z'$. The double cover $\mathrm{Sp}_2(n)$ of the symplectic group plays a very interesting role because it can be realized as a certain group of unitary operators acting on $L^2(\mathbb{R}^n)$. This group is denoted by $\mathrm{Mp}(n)$ and called the *metaplectic group*.

The symplectic group is generated by the set of elementary symplectic matrices

$$M_L = \begin{pmatrix} L^{-1} & 0 \\ 0 & L^T \end{pmatrix}, \quad V_{-P} = \begin{pmatrix} I & 0 \\ P & I \end{pmatrix}, \quad J = \begin{pmatrix} 0 & I \\ -I & 0 \end{pmatrix}, \qquad (5.1)$$

where $L$ ranges over $\mathrm{GL}(n, \mathbb{R})$ and $P = P^T$.

Let us call a symplectic matrix $S \in \mathrm{Sp}(n)$ *free* if it can be written

$$S = \begin{pmatrix} A & B \\ C & D \end{pmatrix} \quad \text{and} \quad \det B \neq 0.$$

It turns out that such a symplectic matrix can always be written in the form $S = V_{-P} M_L J V_{-Q}$, namely

$$S = V_{-DB^{-1}} M_{B^{-1}} J V_{-B^{-1}A}; \qquad (5.2)$$

conversely every symplectic matrix of the type $S = V_{-P} M_L J V_{-Q}$ is free. The remarkable fact is that every symplectic matrix can be written as the *product of exactly two free symplectic matrices*. To every free symplectic matrix we can associated the generating function

$$\mathcal{A}(x, x') = \frac{1}{2} DB^{-1} x^2 - B^{-1} x \cdot x' + \frac{1}{2} B^{-1} A x'^2;$$

the terminology is motivated by the fact that we have

$$(x, p) = S(x', p') \iff p = \partial_x \mathcal{A}(x, x'), p' = -\partial_{x'} \mathcal{A}(x, x').$$

In the metaplectic case, we have the following results: the group $\mathrm{Mp}(n)$ is generated by the unitary operators $\widehat{M}_{L,m}, \widehat{V}_{-P}$, and $\widehat{J}$ defined by

$$\widehat{V}_{-P} \psi(x) = e^{\frac{i}{2\hbar} P x \cdot x} \psi(x), \qquad (5.3)$$

$$\widehat{M}_{L,m} \psi(x) = i^m \sqrt{|\det L|} \psi(Lx), \qquad (5.4)$$

$$\widehat{J} \psi(x) = e^{in\pi/4} F \psi(x). \qquad (5.5)$$

Note that the transformations (5.3) are called "chirps" in the time-frequency analysis literature; in (5.4) the integer $m$ corresponds to a choice of $\arg \det L$ ($m \equiv 0 \bmod 4$ if $\det L > 0$ and $m \equiv 2 \bmod 4$ if $\det L < 0$); in (5.5) $F$ is the usual ($\hbar$-dependent) unitary Fourier transform.

Let $S = V_{-P} M_L J V_{-Q}$ be a free symplectic matrix, and consider the metaplectic operator $\widehat{S} = \widehat{V}_{-P} \widehat{M}_{L,m} \widehat{J} \widehat{V}_{-Q}$; an easy calculation shows that $\widehat{S}$ is given by

$$\widehat{S}\psi(x) = \left(\frac{1}{2\pi\hbar}\right)^{n/2} e^{in\pi/4} \Delta(\mathcal{A}) \int_{\mathbb{R}^n} e^{\frac{i}{\hbar} A(x,x')} \psi(x')dx', \qquad (5.6)$$

where the factor $\Delta(\mathcal{A})$ is defined by

$$\Delta(\mathcal{A}) = i^m \sqrt{|\det L|}. \qquad (5.7)$$

These operators $\widehat{S} = \widehat{S}_{\mathcal{A},m}$ generate $\mathrm{Mp}(n)$; in fact (see [33]) every $\widehat{S} \in \mathrm{Mp}(n)$ can be written as a product of exactly two operators of this type: $\widehat{S} = \widehat{S}_{\mathcal{A},m} \widehat{S}_{\mathcal{A}',m'}$.

### 5.1.2. Proof of the symplectic covariance property

One of the most interesting properties of the (cross-)Wigner transform is the way it behaves under linear symplectic transformations. The following result is fundamental.

**Proposition 57.** *Let $S \in \mathrm{Sp}(n)$ and let $\widehat{S}$ be anyone of the two elements of $\mathrm{Mp}(n)$ covering $S$.*
(i) *For every $z \in \mathbb{R}^{2n}$ we have*

$$\widehat{T}(Sz) = \widehat{S}\widehat{T}(z)\widehat{S}^{-1} \quad \text{and} \quad \widehat{R}(Sz) = \widehat{S}\widehat{R}(z)\widehat{S}^{-1}, \qquad (5.8)$$

*where $\widehat{T}(z_0)$ and $\widehat{R}(z_0)$ are, respectively, the Heisenberg–Weyl and the Grossmann–Royer operator.*
(ii) *Let $\psi$ and $\phi$ be in $\mathcal{S}(\mathbb{R}^n)$. We have*

$$W(\psi, \phi)(S^{-1}z) = W(\widehat{S}\psi, \widehat{S}\phi)(z), \qquad (5.9)$$

$$A(\psi, \phi)(S^{-1}z) = A(\widehat{S}\psi, \widehat{S}\phi)(z), \qquad (5.10)$$

*where $\widehat{S}$ is defined as above. In particular*

$$W\psi(S^{-1}z) = W(\widehat{S}\psi)(z), \quad A\psi(S^{-1}z) = A(\widehat{S}\psi)(z). \qquad (5.11)$$

**Proof.** We begin by showing that the identities (5.9) and (5.10) follow from the formulas (5.8). Recall that, by definition

$$W(\psi, \phi)(z) = \left(\frac{1}{\pi\hbar}\right)^n (\widehat{R}(z)\psi|\phi)_{L^2}.$$

We thus have, assuming that the first formula in (5.8) holds,

$$W(\psi, \phi)(S^{-1}z) = \left(\frac{1}{\pi\hbar}\right)^n (\widehat{R}(S^{-1}z)\psi|\phi)_{L^2}$$

$$= \left(\frac{1}{\pi\hbar}\right)^n (\widehat{S}\widehat{R}(z)\widehat{S}^{-1}\psi|\phi)_{L^2};$$

metaplectic operators being unitary we have

$$(\widehat{S}\widehat{R}(z)\widehat{S}^{-1}\psi|\phi)_{L^2(\mathbb{R}^n)} = (\widehat{R}(z)\widehat{S}^{-1}\psi|\widehat{S}^{-1}\phi)_{L^2}$$

and hence

$$W(\psi, \phi)(S^{-1}z) = \left(\frac{1}{\pi\hbar}\right)^n (\widehat{R}(z)\widehat{S}^{-1}\psi|\widehat{S}^{-1}\phi)_{L^2}$$

$$= W(\widehat{S}\psi, \widehat{S}\phi)(z).$$

That (5.10) follows from the second formula (5.8) is proven exactly in the same way, using the definition

$$A(\psi, \phi)(z) = \left(\frac{1}{2\pi\hbar}\right)^n (\psi|\widehat{T}(z)\phi)_{L^2}$$

of the cross-ambiguity function. Let us now prove the identities (5.8). Let us first prove that

$$\widehat{T}(Sz) = \widehat{S}\widehat{T}(z)\widehat{S}^{-1}. \tag{5.12}$$

For this it is no restriction to assume that $\widehat{S}$ is of the type $\widehat{S}_{A,m}$: since every $\widehat{S} \in \mathrm{Mp}(n)$ is a product $\widehat{S} = \widehat{S}_{A,m}\widehat{S}_{A',m'}$ of two such operators, we have then

$$\widehat{T}(Sz) = \widehat{S}_{A,m}(\widehat{S}_{A',m'}\widehat{T}(z)\widehat{S}_{A',m'}^{-1})\widehat{S}_{A,m}^{-1}$$

$$= \widehat{S}_{A,m}\widehat{T}(S_{A'}z)\widehat{S}_{A,m}^{-1}$$

$$= \widehat{T}(S_A S_{A'} z)$$

$$= \widehat{S}_{A,m} \widehat{S}_{A',m'} \widehat{T}(z)(\widehat{S}_{A,m} \widehat{S}_{A',m'})^{-1}$$

$$= \widehat{S}\widehat{T}(z)\widehat{S}^{-1}.$$

Let us thus prove (5.12) for $\widehat{S} = \widehat{S}_{A,m}$; equivalently, we have to show that

$$\widehat{T}(z_0)\widehat{S}_{A,m} = \widehat{S}_{A,m}\widehat{T}(S_A^{-1} z_0) \tag{5.13}$$

(we change $z$ in $z_0$ to avoid notational confusion when the variable $x$ becomes involved). For $\psi \in \mathcal{S}(\mathbb{R}^n)$ set

$$g(x) = \widehat{T}(z_0)\widehat{S}_{A,m}\psi(x).$$

By definition (A.15) of $\widehat{S}_{A,m}$ and definition (1.8) of the Heisenberg–Weyl operator $\widehat{T}(z_0)$ we have

$$g(x) = e^{\frac{i}{\hbar}(p_0 \cdot x - \frac{1}{2}p_0 \cdot x_0)}\widehat{S}_{A,m}\psi(x - x_0)$$

$$\times \left(\frac{1}{2\pi i\hbar}\right)^{n/2} \Delta(A)e^{-\frac{1}{2\hbar}p_0 \cdot x_0} \int_{\mathbb{R}^n} e^{\frac{i}{\hbar}(A(x - x_0, x') + p_0 \cdot x)}\psi(x')dx'.$$

In view of formula (A.24) in Proposition A.8, the function

$$A_0(x, x') = A(x - x_0, x') + p_0 \cdot x \tag{5.14}$$

is a generating function of the free affine symplectomorphism $T(z_0)S_A$, hence we have just shown that

$$\widehat{T}(z_0)\widehat{S}_{A,m} = e^{\frac{i}{2\hbar}p_0 \cdot x_0}\widehat{S}_{A_0,m}, \tag{5.15}$$

where $\widehat{S}_{A_0,m}$ is one of the metaplectic operators associated to $A_0$. Let us now set

$$h(x) = \widehat{S}_{A,m}\widehat{T}(S_A^{-1} z_0)\psi(x) \text{ and } (x_0', p_0') = \widehat{S}_{A,m}^{-1}(x_0, p_0).$$

We have

$$h(x) = \left(\frac{1}{2\pi i\hbar}\right)^{n/2} \Delta(A) \int_{\mathbb{R}^n} e^{\frac{i}{\hbar}A(x, x')}e^{-\frac{i}{2\hbar}p_0' \cdot x_0'}e^{\frac{i}{\hbar}p_0' \cdot x'}\psi(x' - x_0')\,dx',$$

i.e. performing the change of variables $x' \longmapsto x' + x_0'$ such that

$$h(x) = \left(\frac{1}{2\pi i\hbar}\right)^{n/2} \Delta(\mathcal{A}) \int_{\mathbb{R}^n} e^{\frac{i}{\hbar}\mathcal{A}(x,x'+x_0')} e^{\frac{i}{2\hbar}p_0' \cdot x_0'} e^{\frac{i}{\hbar}p_0' \cdot x'} \psi(x') \, dx'.$$

We will thus have $h(x) = g(x)$ as claimed, if we show that

$$\mathcal{A}(x, x' + x_0') + \frac{1}{2}p_0' \cdot x_0' + p_0' \cdot x' = \mathcal{A}_0(x, x') - \frac{1}{2}p_0 \cdot x_0,$$

i.e.

$$\mathcal{A}(x, x' + x_0') + \frac{1}{2}p_0' \cdot x_0' + p_0' \cdot x' = \mathcal{A}(x - x_0, x') + p_0 \cdot x - \frac{1}{2}p_0 \cdot x_0.$$

Replacing $x$ by $x + x_0$ this amounts to prove that

$$\mathcal{A}(x + x_0, x' + x_0') + \frac{1}{2}p_0' \cdot x_0' + p_0' \cdot x' = \mathcal{A}(x, x') + \frac{1}{2}p_0 \cdot x_0 + p_0 \cdot x.$$

But this equality immediately follows from Proposition A.8 in Appendix A and its Corollary A.9.                                                     □

### 5.1.3.  Symplectic covariance of Weyl operators

We have proven in Proposition 57 above the following symplectic covariance property for the cross-Wigner transform: let $S \in \mathrm{Sp}(n)$ and $\widehat{S} \in \mathrm{Mp}(n)$ covering $S$. We have

$$W(\psi, \phi)(S^{-1}z) = W(\widehat{S}\psi, \widehat{S}\phi)(z). \tag{5.16}$$

We are going to use this result to prove a characteristic property of the Weyl correspondence.

**Proposition 58.** *Let $a \in \mathcal{S}'(\mathbb{R}^{2n})$ and $\widehat{A} = \mathrm{Op_W}(a)$.*

(i) *Let $S \in \mathrm{Sp}(n)$ and $\widehat{S} \in \mathrm{Mp}(n)$ be anyone of the two metaplectic operators covering $S$. We have*

$$\mathrm{Op_W}(a \circ S) = \widehat{S}^{-1}\mathrm{Op_W}(a)\widehat{S}. \tag{5.17}$$

(ii) *Let $T(z_0)a(z) = a(z - z_0)$. We have*

$$\mathrm{Op_W}(T(z_0)a) = \widehat{T}(z_0)\mathrm{Op_W}(a)\widehat{T}(z_0)^{-1}. \tag{5.18}$$

**Proof.** Let $\psi, \phi \in \mathcal{S}(\mathbb{R}^n)$.

(i) In view of formula (4.17) in Proposition 49 we have

$$(\mathrm{Op_W}(a \circ S)\psi|\phi)_{L^2} = \int_{\mathbb{R}^{2n}} a(Sz)W(\psi, \phi)(z)dz$$

$$= \int_{\mathbb{R}^{2n}} a(z)W(\psi, \phi)(S^{-1}z)dz.$$

Using (5.16) this equality becomes

$$(\mathrm{Op_W}(a \circ S)\psi|\phi)_{L^2} = \int_{\mathbb{R}^{2n}} a(z)W(\widehat{S}\psi, \widehat{S}\phi)(z)dz$$

$$= (\mathrm{Op_W}(a)(\widehat{S}\psi)|\widehat{S}\phi)_{L^2}$$

$$= (\widehat{S}^{-1}\mathrm{Op_W}(a))(\widehat{S}\psi)|\phi)_{L^2},$$

which proves formula (5.17) for $\psi, \phi \in \mathcal{S}(\mathbb{R}^n)$. The general case follows by a continuity and density argument.

(ii) In view of the translation formula (2.30) in Chapter 2, we have

$$W(\widehat{T}(z_0)^{-1}\phi, \widehat{T}(z_0)^{-1}\psi) = W(\phi, \psi)T(z_0)^{-1}$$

and hence, using again formula (4.17),

$$\langle \widehat{T}(z_0)\widehat{A}\widehat{T}(z_0)^{-1}\psi, \phi \rangle = \langle \widehat{A}\widehat{T}(z_0)^{-1}\psi, \widehat{T}(z_0)^{-1}\phi \rangle$$

$$= \langle a, W(\widehat{T}(z_0)^{-1}\phi, \widehat{T}(z_0)^{-1}\psi) \rangle$$

$$= \langle a, T(-z_0)W(\phi, \psi) \rangle.$$

Since we obviously have

$$\langle a, T(-z_0)W(\phi, \psi) \rangle = \langle T(z_0)a, W(\phi, \psi) \rangle,$$

formula (5.18) follows. $\qquad \square$

## 5.2. Maximal Covariance

It turns out that we cannot extend very much the symplectic covariance results in Propositions 57(ii) and 58. We begin by studying antisymplectic matrices. We are following closely the exposition we have given in [18] with Dias and Prata.

### 5.2.1. Antisymplectic matrices

We have seen that the cross-Wigner transform and the cross-ambiguity function enjoy covariance with respect to linear symplectic transformations. A natural question to ask in this context is whether we could in some way improve Proposition 57 by, for instance, extending in some way the groups $\mathrm{Sp}(n)$ and $\mathrm{Mp}(n)$. We are going to show that this is not possible, and that symplectic covariance is in a sense an optimal symmetry property.

**Definition 59.** An automorphism $M$ of $\mathbb{R}^{2n}$ is antisymplectic if $\sigma(Mz, Mz') = -\sigma(z, z')$ for all $z, z' \in \mathbb{R}^{2n}$; in matrix notation $M^T J M = -J$. Equivalently $CM \in \mathrm{Sp}(n)$ where $C = \begin{pmatrix} I & 0 \\ 0 & -I \end{pmatrix}$.

**Notation 60.** We denote by $\mathrm{Sp}^+(n)$ the subset of $\mathrm{Sp}(n)$ consisting of symmetric positive definite symplectic matrices. If $G \in \mathrm{Sp}^+(n)$ then $G^\alpha \in \mathrm{Sp}^+(n)$ for every $\alpha \in \mathbb{R}$.

We will need the following algebraic lemma.

**Lemma 61.** *Let $M \in \mathrm{GL}(2n, \mathbb{R})$ and assume that $M^T G M \in \mathrm{Sp}(n)$ for every for every block-diagonal $G \in \mathrm{Sp}^+(n)$:*

$$G = \begin{pmatrix} X & 0 \\ 0 & X^{-1} \end{pmatrix} \in \mathrm{Sp}^+(n). \tag{5.19}$$

*Then $M$ is either symplectic, or antisymplectic.*

**Proof.** We first remark that, taking $G = I_\mathrm{d}$ in the condition $M^T G M \in \mathrm{Sp}(n)$, we have $M^T M \in \mathrm{Sp}(n)$. Next, we can write $M = HP$ where $H = M(M^T M)^{-1/2}$ is orthogonal and $P = (M^T M)^{1/2} \in \mathrm{Sp}^+(n)$ (this is easily seen using the polar decomposition theorem). It follows that the condition $M^T G M \in \mathrm{Sp}(n)$ is equivalent to $P(H^T GH)P \in \mathrm{Sp}(n)$; since $P$ is symplectic so is $P^{-1}$ and hence $H^T GH \in \mathrm{Sp}(n)$ for all $G$ of the form (5.19). Let us now choose $G$ diagonal:

$$G = \begin{pmatrix} \Lambda & 0 \\ 0 & \Lambda^{-1} \end{pmatrix},$$

where $\Lambda$ is the diagonal matrix with diagonal elements $\lambda_1, \ldots, \lambda_n$ where $\lambda_j > 0$ for $1 \le j \le n$. We have

$$H^T \begin{pmatrix} \Lambda & 0 \\ 0 & \Lambda^{-1} \end{pmatrix} H \in \mathrm{Sp}^+(n)$$

for every $\Lambda$ of this type. Let

$$U \in U(n) = \mathrm{Sp}(n) \cap O(2n, \mathbb{R})$$

be such that

$$H^T \begin{pmatrix} \Lambda & 0 \\ 0 & \Lambda^{-1} \end{pmatrix} H = U^T \begin{pmatrix} \Lambda & 0 \\ 0 & \Lambda^{-1} \end{pmatrix} U;$$

we have $H^T G H \in \mathrm{Sp}^+(n)$ and the eigenvalues of $H^T G H$ are those of $G$ since $H$ is orthogonal. Setting $R = HU^T$ the equality above is equivalent to

$$\begin{pmatrix} \Lambda & 0 \\ 0 & \Lambda^{-1} \end{pmatrix} R = R \begin{pmatrix} \Lambda & 0 \\ 0 & \Lambda^{-1} \end{pmatrix}. \tag{5.20}$$

Writing $R = \begin{pmatrix} A & B \\ C & D \end{pmatrix}$ we get the conditions

$$\Lambda A = A\Lambda, \ \Lambda B = B\Lambda^{-1},$$

$$\Lambda^{-1} C = C\Lambda, \ \Lambda^{-1} D = D\Lambda^{-1}$$

for all $\Lambda$. It follows from these conditions that $A$ and $D$ must themselves be diagonal:

$$A = \begin{pmatrix} a_1 & \cdots & 0 \\ 0 & \ddots & 0 \\ 0 & \cdots & a_n \end{pmatrix}, \quad D = \begin{pmatrix} d_1 & \cdots & 0 \\ 0 & \ddots & 0 \\ 0 & \cdots & d_n \end{pmatrix}.$$

On the other hand, choosing $\Lambda = \lambda I$, $\lambda \neq 1$, we get $B = C = 0$. Hence, taking into account the fact that $R \in O(2n, \mathbb{R})$ we must have

$$R = \begin{pmatrix} A & 0 \\ 0 & D \end{pmatrix}, \quad A^2 = D^2 = I. \tag{5.21}$$

Conversely, if $R$ is of the form (5.21), then (5.20) holds for any positive-definite diagonal $\Lambda$. We conclude that $M$ has to be of the form $M = RUP$ where $R$ is of the form (5.21). Since $UP \in \mathrm{Sp}(n)$, $M^T GM \in \mathrm{Sp}(n)$ implies $RGR \in \mathrm{Sp}(n)$. We now choose, for $1 \leq i < j \leq n$

$$X^{(ij)} = I + \frac{1}{2} E^{(ij)}, \tag{5.22}$$

where $E^{(ij)}$ is the symmetric matrix whose entries are all zero except the ones on the $i$th row and $j$th column and on the $j$th row and $i$th column which are equal to one. A simple calculation then shows that

$$RGR = \begin{pmatrix} AX^{(ij)}A & 0 \\ 0 & D(X^{(ij)})^{-1}D \end{pmatrix}. \qquad (5.23)$$

The condition $RGR \in \mathrm{Sp}(n)$ implies that

$$AX^{(ij)}AD(X^{(ij)})^{-1}D = I \iff X^{(ij)}AD = ADX^{(ij)}; \qquad (5.24)$$

the matrix $AD$ thus commutes with every real positive-definite $n \times n$ matrix $X^{(ij)}$ of the form (5.22). Writing $AD = \mathrm{diag}(c_1, \ldots, c_n)$ with $c_j = a_j d_j$ for $1 \leq j \leq n$ and applying (5.24) to (5.22) for $i < j$, we conclude that $c_i = c_j$. This implies that the entries of the matrix $AD$ are all equal, that is, either $AD = I$ or $AD = -I$, or equivalently $A = D$ or $A = -D$. In the first case, $R$ is symplectic and so is $M$. In the second case $R$ is antisymplectic; but then $M$ is also antisymplectic. $\qquad \square$

### 5.2.2. The maximality property

We can now prove the main result of this section.

**Proposition 62.** *Let $M \in \mathrm{GL}(2n, \mathbb{R})$.*

(i) *Assume that $M$ is antisymplectic: $S = CM \in \mathrm{Sp}(n)$ where $C = \begin{pmatrix} I & 0 \\ 0 & -I \end{pmatrix}$; then for every $\psi \in \mathcal{S}'(\mathbb{R}^n)$ we have*

$$W\psi(Mz) = W(\widehat{S}^{-1}\overline{\psi})(z), \qquad (5.25)$$

*where $\widehat{S}$ is any of the two elements of $\mathrm{Mp}(n)$ covering $S$.*

(ii) *Conversely, assume that for any $\psi \in \mathcal{S}(\mathbb{R}^n)$ there exists $\psi' \in \mathcal{S}'(\mathbb{R}^n)$ such that*

$$W\psi(Mz) = W\psi'(z). \qquad (5.26)$$

*Then $M$ is either symplectic or antisymplectic.*

**Proof.** (i) It is sufficient to assume that $\psi \in \mathcal{S}(\mathbb{R}^n)$. We have

$$W\psi(Cz) = \left(\frac{1}{2\pi\hbar}\right)^n \int_{\mathbb{R}^n} e^{\frac{i}{\hbar}p \cdot y}\psi\left(x + \frac{1}{2}y\right)\overline{\psi\left(x - \frac{1}{2}y\right)}dy$$

$$= \left(\frac{1}{2\pi\hbar}\right)^n \int_{\mathbb{R}^n} e^{-\frac{i}{\hbar}p \cdot y}\psi\left(x - \frac{1}{2}y\right)\overline{\psi\left(x + \frac{1}{2}y\right)}dy$$

$$= W\overline{\psi}(z).$$

It follows that

$$W\psi(Mz) = W\psi(CSz) = W\overline{\psi}(Sz)$$

hence formula (5.25).

(ii) Choosing for $\psi$ a normalized Gaussian of the form

$$\psi_X(x) = \left(\frac{1}{\pi\hbar}\right)^{n/4} (\det X)^{1/4} e^{-\frac{1}{2\hbar}Xx\cdot x} \tag{5.27}$$

($X$ is real symmetric and positive definite), we have (see formula (9.13) in Corollary 113, Chapter 9)

$$W\psi_X(z) = \left(\frac{1}{\pi\hbar}\right)^n e^{-\frac{1}{\hbar}Gz\cdot z}, \tag{5.28}$$

where

$$G = \begin{pmatrix} X & 0 \\ 0 & X^{-1} \end{pmatrix} \tag{5.29}$$

is positive definite and belongs to Sp($n$): $G \in$ Sp$^+(n)$. Condition (5.26) implies that we must have

$$W\psi'(z) = \left(\frac{1}{\pi\hbar}\right)^n e^{-\frac{1}{\hbar}M^T GMz\cdot z}.$$

The Wigner transform of a function being a Gaussian if and only if the function itself is a Gaussian (see Chapter 9). This can be seen in the following way: the matrix $M^T GM$ being symmetric and positive definite and we can use a Williamson diagonalization (see Appendix C): there exists $S \in$ Sp($n$) such that

$$S^T(M^T GM)S = \Delta = \begin{pmatrix} \Sigma & 0 \\ 0 & \Sigma \end{pmatrix} \tag{5.30}$$

and hence

$$W\psi'(Sz) = \left(\frac{1}{\pi\hbar}\right)^n e^{-\frac{1}{\hbar}\Delta z\cdot z}.$$

In view of the symplectic covariance of the Wigner transform, we have

$$W\psi'(Sz) = W\psi''(z), \quad \psi'' = \widehat{S}^{-1}\psi'$$

where $\widehat{S} \in$ Mp($n$) is one of the two elements of the metaplectic group covering $S$. We now show that the equality

$$W\psi''(z) = \left(\frac{1}{\pi\hbar}\right)^n e^{-\frac{1}{\hbar}\Delta z\cdot z} \tag{5.31}$$

implies that $\psi''$ must be a Gaussian of the form (5.27) and hence $W\psi''$ must be of the type (5.28)–(5.29). That $\psi''$ must be a Gaussian follows from $W\psi'' \geq 0$ and Hudson's theorem (see Chapter 9). If $\psi''$ were of the more general type

$$\psi_{X,Y}(x) = \left(\frac{1}{\pi\hbar}\right)^{n/4} (\det X)^{1/4} e^{-\frac{1}{2\hbar}(X+iY)x \cdot x} \tag{5.32}$$

($X, Y$ are real and symmetric and $X$ is positive definite) the matrix $G$ in (5.28) would be

$$G = \begin{pmatrix} X + YX^{-1}Y & YX^{-1} \\ X^{-1}Y & X^{-1} \end{pmatrix}$$

which is only compatible with (5.31) if $Y = 0$. In addition, due to the parity of $W\psi''$, the function $\psi''$ must be an even function, hence Gaussians more general than $\psi_{X,Y}$ are excluded. It follows from these considerations that we have

$$\Delta = \begin{pmatrix} \Sigma & 0 \\ 0 & \Sigma \end{pmatrix} = \begin{pmatrix} X & 0 \\ 0 & X^{-1} \end{pmatrix}$$

so that $\Sigma = \Sigma^{-1}$. Since $\Sigma > 0$ this implies that we must have $\Sigma = I$, and hence, using formula (5.30), $S^T(M^TGM)S = I$. It follows that we must have $M^TGM \in \mathrm{Sp}(n)$ for every $G = \begin{pmatrix} X & 0 \\ 0 & X^{-1} \end{pmatrix} \in \mathrm{Sp}^+(n)$. In view of Lemma 61, the matrix $M$ must then be either symplectic or antisymplectic. $\qquad\square$

### 5.2.3. The case of Weyl operators

Proposition 58 where we established the symplectic covariance of Weyl operators is optimal as well, that is, one cannot expect it to hold for larger groups of operators.

**Proposition 63.** *Let* $M \in \mathrm{GL}(2n, \mathbb{R})$. *Assume that there exists a unitary operator* $\widehat{M} : L^2(\mathbb{R}^n) \longrightarrow L^2(\mathbb{R}^n)$ *such that*

$$\mathrm{Op_W}(a \circ M) = \widehat{M}^{-1}\mathrm{Op_W}(a)\widehat{M} \tag{5.33}$$

*for all* $\widehat{A} = \mathrm{Op_W}(a)$, $a \in \mathcal{S}(\mathbb{R}^{2n})$. *Then either* $M \in \mathrm{Sp}(n)$ *or* $M$ *is antisymplectic.*

**Proof.** Suppose that (5.33) holds; then

$$(\widehat{M}^{-1}\widehat{A}\widehat{M}\psi|\phi)_{L^2} = \langle a \circ M, W(\psi,\phi)\rangle = \langle a, W(\psi,\phi) \circ M\rangle.$$

On the other hand, since $\widehat{M}$ is unitary,

$$(\widehat{M}^{-1}\widehat{A}\widehat{M}\psi|\phi)_{L^2} = (\widehat{A}\widehat{M}\psi|\widehat{M}\phi)_{L^2} = \langle a, W(\widehat{M}\psi,\widehat{M}\phi)\rangle.$$

It follows that we must have

$$\langle a, W(\psi,\phi) \circ M^{-1}\rangle = \langle a, W(\widehat{M}\psi,\widehat{M}\phi)\rangle$$

for all $\psi, \phi \in \mathcal{S}(\mathbb{R}^n)$ and hence, in particular, taking $\psi = \phi$:

$$\langle a, W\psi \circ M^{-1}\rangle = \langle a, W(\widehat{M}\psi)\rangle$$

for all $\psi \in \mathcal{S}(\mathbb{R}^n)$. Since $a$ is arbitrary, this implies that we must have $W\psi \circ M = W(\widehat{M}^{-1}\psi)$. In view of Proposition 62, the automorphism $M$ must be either symplectic or antisymplectic. $\qquad\square$

## Main Formulas in Chapter 5

| | |
|---|---|
| Generator of Mp($n$) | $\widehat{S}_{W,m}\psi(x) = \left(\frac{1}{2\pi\hbar i}\right)^{n/2}\Delta(W)$ |
| | $\int_{\mathbb{R}^n} e^{\frac{i}{\hbar}W(x,x')}\psi(x')dx'$ |
| Generating function | $W(x,x') = \frac{1}{2}DB^{-1}x^2 - B^{-1}x \cdot x'$ |
| | $+\frac{1}{2}B^{-1}Ax'^2$ |
| Factor $\Delta(W)$ | $\Delta(W) = i^m\sqrt{|\det L|}, \quad m\pi = \arg \det L$ |
| Generator of Sp($n$) | $S_W = \begin{pmatrix} A & B \\ C & D \end{pmatrix}, \det B \neq 0$ |
| Symplectic covariance (HW/GR) | $\widehat{T}(Sz) = \widehat{S}\widehat{T}(z)\widehat{S}^{-1},$ |
| | $\widehat{R}(Sz) = \widehat{S}\widehat{R}(z)\widehat{S}^{-1}$ |
| Symplectic covariance (CWT) | $W(\psi,\phi)(S^{-1}z) = W(\widehat{S}\psi,\widehat{S}\phi)(z)$ |
| Symplectic covariance (CAF) | $A(\psi,\phi)(S^{-1}z) = A(\widehat{S}\psi,\widehat{S}\phi)(z)$ |
| Maximal covariance | $W\psi(Mz) = W\psi'(z) \Longrightarrow MJM^T = \pm J$ |

# Chapter 6

# The Moyal Identity

**Summary 64.** The Moyal identity in its simplest form shows that the $L^2$-norms of $W\psi$ and $\psi$ are equal up to a constant factor. It is an essential tool for the proof of the reconstruction formula.

The Moyal identity for the cross-Wigner transform is an extremely useful tool in harmonic analysis and quantum mechanics. It is also essential for the reconstruction formulas for the cross-Wigner transform and its twin, the cross-ambiguity function. These results are essential in the study the notion of weak value in quantum mechanics.

## 6.1. Precise Statement and Proof

## 6.1.1. The general Moyal identity

The main result of this section is the following proposition.

**Proposition 65.** *The cross-Wigner transform satisfies the "Moyal identity"*

$$(W(\psi,\phi)|W(\psi',\phi'))_{L^2(\mathbb{R}^{2n})} = \left(\frac{1}{2\pi\hbar}\right)^n (\psi|\psi')_{L^2}\overline{(\phi|\phi')_{L^2}} \tag{6.1}$$

*for all $(\psi,\phi) \in L^2(\mathbb{R}^n) \times L^2(\mathbb{R}^n)$. In particular*

$$||W\psi||_{L^2(\mathbb{R}^{2n})} = \left(\frac{1}{2\pi\hbar}\right)^{n/2} ||\psi||_{L^2}. \tag{6.2}$$

**Proof.** Let us set

$$A = (2\pi\hbar)^{2n} (W(\psi, \phi) | W(\psi', \phi'))_{L^2(\mathbb{R}^{2n})}.$$

In view of the integral definition (2.5) of the cross-Wigner transform, we have

$$A = \int_{\mathbb{R}^{4n}} e^{-\frac{i}{\hbar} p \cdot (y - y')} \psi \left( x + \frac{1}{2} y \right) \overline{\psi' \left( x + \frac{1}{2} y' \right)}$$

$$\times \overline{\phi \left( x - \frac{1}{2} y \right)} \phi' \left( x - \frac{1}{2} y' \right) dy \, dy' \, dx \, dp.$$

The integral in $p$ can be viewed as an inverse Fourier transform, yielding

$$\int_{\mathbb{R}^n} e^{-\frac{i}{\hbar} p \cdot (y - y')} dp = (2\pi\hbar)^n \delta(y - y')$$

and hence

$$A = (2\pi\hbar)^n \int_{\mathbb{R}^{3n}} \psi \left( x + \frac{1}{2} y \right) \overline{\psi' \left( x - \frac{1}{2} y \right)} \phi \left( x + \frac{1}{2} y \right) \overline{\phi'}$$

$$\times \left( x - \frac{1}{2} y \right) dy \, dy' \, dx.$$

Setting $x' = x + \frac{1}{2} y$ and $y' = x - \frac{1}{2} y$ we have $dx' dy' = dx dy$ and hence

$$A = (2\pi\hbar)^n \left( \int_{\mathbb{R}^n} \psi(x') \overline{\psi'(x')} dx' \right) \left( \int_{\mathbb{R}^n} \overline{\phi(y')} \phi'(y') dy' \right)$$

which proves Moyal's identity (6.1). Therefore formula (6.2) immediately follows.                                                                    □

**Remark 66.** One could also prove the Moyal identity by using the factorization result (2.15) in Proposition 16.

Notice that the Moyal identity can be written, using distributional brackets, as

$$\langle W(\psi, \phi), W(\psi', \phi') \rangle = \left( \frac{1}{2\pi\hbar} \right)^n \langle \psi, \overline{\psi'} \rangle \langle \phi, \overline{\phi'} \rangle. \tag{6.3}$$

An immediate consequence of Proposition 65 is that the cross-ambiguity transform satisfies a similar identity.

**Corollary 67.** *The cross-ambiguity function satisfies the Moyal identity*

$$(A(\psi,\phi)|A(\psi',\phi'))_{L^2(\mathbb{R}^{2n})} = \left(\frac{1}{2\pi\hbar}\right)^n (\psi|\psi')_{L^2}\overline{(\phi|\phi')_{L^2}} \qquad (6.4)$$

*for all $\psi,\phi \in L^2(\mathbb{R}^n)$. In particular*

$$\|A\psi\|_{L^2(\mathbb{R}^{2n})} = \left(\frac{1}{2\pi\hbar}\right)^{n/2} \|\psi\|_{L^2}. \qquad (6.5)$$

**Proof.** In view of Proposition 65, we have

$$(W(\psi,\phi)|W(\psi',\phi'))_{L^2(\mathbb{R}^{2n})} = \left(\frac{1}{2\pi\hbar}\right)^n (\psi|\psi')_{L^2}\overline{(\phi|\phi')_{L^2}};$$

formula (6.4) follows from the relation $W(\psi,\phi) = F_\sigma A(\psi,\phi)$ and the fact that $F_\sigma$ is a unitary operator on $L^2(\mathbb{R}^{2n})$. □

It trivially follows that the Fourier–Wigner transform (3.11) also satisfies the Moyal identity:

$$(V(\psi,\phi)|V(\psi',\phi'))_{L^2(\mathbb{R}^{2n})} = \left(\frac{1}{2\pi\hbar}\right)^n (\psi|\psi')_{L^2}\overline{(\phi|\phi')_{L^2}} \qquad (6.6)$$

for all $\psi,\phi \in L^2(\mathbb{R}^n)$.

### 6.1.2. A continuity result

The Moyal identity can be used to establish continuity results for the cross-Wigner transform and the cross-ambiguity function.

**Proposition 68.** *The cross-Wigner transform and the cross-ambiguity function are continuous mappings*

$$W : L^2(\mathbb{R}^n) \times L^2(\mathbb{R}^n) \longrightarrow L^2(\mathbb{R}^{2n}), \qquad (6.7)$$

$$A : L^2(\mathbb{R}^n) \times L^2(\mathbb{R}^n) \longrightarrow L^2(\mathbb{R}^{2n}). \qquad (6.8)$$

**Proof.** Choosing $\psi = \psi'$ and $\phi = \phi'$ in Moyal's identity (6.1) we get

$$\|W(\psi,\phi)\|_{L^2(\mathbb{R}^{2n})} = \left(\frac{1}{2\pi\hbar}\right)^{n/2} \|\psi\|_{L^2}\|\phi\|_{L^2}$$

hence the sesquilinear mapping $(\psi,\phi) \longmapsto W(\psi,\phi)$ is a continuous mapping $L^2(\mathbb{R}^n) \times L^2(\mathbb{R}^n) \longrightarrow L^2(\mathbb{R}^{2n})$. The continuity of the cross-ambiguity function $(\psi, \phi) \longmapsto A(\psi, \phi)$ follows: since $A(\psi, \phi)$ is the symplectic Fourier transform of $W(\psi, \phi)$, we have

$$||A(\psi, \phi)||_{L^2(\mathbb{R}^{2n\parallel})} = ||F_\sigma W(\psi, \phi)||_{L^2(\mathbb{R}^{2n\parallel})}$$

$$= \left(\frac{1}{2\pi\hbar}\right)^{n/2} ||\psi||_{L^2|}||\phi||_{L^2}.$$

$\square$

## 6.2. Reconstruction Formulas

We are going to see that the knowledge of the cross-Wigner transform $W(\psi, \phi)$ allows us to reconstruct *both* the functions $\psi$ and $\phi$. This is a truly remarkable result. We will need it when we study the notion of weak value in Chapter 12 and for the definition of Feichtinger's algebra in Chapter 7.

### 6.2.1. Reconstruction using the cross-Wigner transform

Here is a very general reconstruction result.

**Proposition 69.** *Let* $\psi, \phi \in L^2(\mathbb{R}^n)$, $\phi \neq 0$. *We have*

$$\psi(x) = \frac{2^n}{(\phi|\gamma)_{L^2}} \int_{\mathbb{R}^{2n}} W(\psi, \phi)(z_0)\widehat{R}(z_0)\gamma(x)dz_0 \qquad (6.9)$$

*(almost everywhere) for every* $\gamma \in L^2(\mathbb{R}^n)$ *such that* $(\phi|\gamma)_{L^2} \neq 0$. *In particular,*

$$\psi(x) = \frac{2^n}{||\phi||_{L^2}} \int_{\mathbb{R}^{2n}} W(\psi, \phi)(z_0)\widehat{R}(z_0)\phi(x)dz_0. \qquad (6.10)$$

**Proof.** Assume first $\gamma \in \mathcal{S}(\mathbb{R}^n)$. Let us denote by $f(x)$ the integral in the right-hand side of (6.9):

$$f(x) = \int_{\mathbb{R}^{2n}} W(\psi, \phi)(z_0)\widehat{R}(z_0)\gamma(x)dz_0. \qquad (6.11)$$

This integral defines an element of $L^2(\mathbb{R}^n)$; in fact for every $g \in L^2(\mathbb{R}^n)$ we have

$$(f|g)_{L^2} = \int_{\mathbb{R}^{2n}} W(\psi, \phi)(z_0)(\widehat{R}(z_0)\gamma|g)_{L^2}dz_0,$$

i.e.

$$(f|g)_{L^2} = (\pi\hbar)^n \int_{\mathbb{R}^{2n}} W(\psi, \phi)(z_0)W(\gamma, g)(z_0)dz_0. \qquad (6.12)$$

Both $W(\psi, \phi)$ and $W(\gamma, g)$ are in $L^2(\mathbb{R}^{2n})$ hence, using the Cauchy–Schwarz inequality and the conjugation identity $W(\gamma, g) = \overline{W(g, \gamma)}$,

$$|(f|g)_{L^2}|^2 = (\pi\hbar)^{2n}|(W(\psi, \phi)|W(g, \gamma))_{L^2(\mathbb{R}^{2n})}|^2 < \infty.$$

On the other hand, applying the Moyal identity to the equality (6.12) we get

$$(f|g)_{L^2} = 2^{-n}(\psi|g)_{L^2}(\phi|\gamma)_{L^2};$$

since this equality holds for all $g \in L^2(\mathbb{R}^n)$ we must have

$$f(x) = 2^{-n}(\phi|\gamma)_{L^2}\psi(x)$$

almost everywhere. It follows, returning to the definition of $f$, that

$$2^{-n}(\phi|\gamma)_{L^2}\psi(x) = \int_{\mathbb{R}^{2n}} W(\psi, \phi)(z_0)\widehat{R}(z_0)\gamma(x)dz_0$$

(almost everywhere), that is

$$\psi(x) = \frac{2^n}{(\phi|\gamma)_{L^2}} \int_{\mathbb{R}^{2n}} W(\psi, \phi)(z_0)\widehat{R}(z_0)\gamma(x)dz_0,$$

which we set out to prove. □

### 6.2.2. Reconstruction using the cross-ambiguity function

We are going to prove a formula similar to (6.9), but involving the cross-ambiguity function rather than the cross-Wigner transform. We could proceed exactly as in the proof of Proposition 69. However, using the fact that $W(\psi, \phi)$ and $A(\psi, \phi)$ are symplectic Fourier transforms of each other, the result is quite straightforward.

**Proposition 70.** *Let* $\psi, \phi \in L^2(\mathbb{R}^n)$, $\phi \neq 0$. *We have*

$$\psi(x) = \frac{1}{(\phi|\gamma)_{L^2}} \int_{\mathbb{R}^{2n}} A(\psi, \phi)(z_0)\widehat{T}(z_0)\gamma(x)dz_0 \qquad (6.13)$$

*(almost everywhere) for every* $\gamma \in L^2(\mathbb{R}^n)$ *such that* $(\phi|\gamma)_{L^2} \neq 0$. *In particular,*

$$\psi(x) = \frac{1}{||\phi||_{L^2}} \int_{\mathbb{R}^{2n}} A(\psi, \phi)(z_0)\widehat{T}(z_0)\phi(x)dz_0. \qquad (6.14)$$

**Proof.** Let us write

$$g(x) = \int_{\mathbb{R}^{2n}} A(\psi, \phi)(z_0)\widehat{T}(z_0)\gamma(x)dz_0.$$

Using the Plancherel formula (B.9) in Appendix B for the symplectic Fourier transform $F_\sigma$, we have

$$g(x) = \int_{\mathbb{R}^{2n}} F_\sigma A(\psi, \phi)(z_0) F_\sigma[\widehat{T}(\cdot)\gamma(x)](-z_0)dz_0,$$

that is, since $F_\sigma A(\psi, \phi) = W(\psi, \phi)$ and using formula (1.20) in Chapter 1,

$$g(x) = 2^n \int_{\mathbb{R}^{2n}} W(\psi, \phi)(z_0)\widehat{R}(z_0)\gamma(x)dz_0.$$

It follows from the proof of Proposition 69 that we have

$$g(x) = (\phi|\gamma)_{L^2}\psi(x).$$

Therefore, formulas (6.13) and (6.14) follow. □

## 6.3. The Wavepacket Transforms

### 6.3.1. Definition

The wavepacket transforms $U_\phi$ are partial isometries from $L^2(\mathbb{R}^n)$ into $L^2(\mathbb{R}^{2n})$ defined using the cross-Wigner transform $W(\psi, \phi)$. They have a number of interesting properties. We will use them in Chapter 11 as "intertwiners" between ordinary (configuration space) quantum mechanics and phase space quantum mechanics (deformation quantization).

**Definition 71.** Let $\phi \in \mathcal{S}(\mathbb{R}^n)$ be normalized in the $L^2$-norm: $||\phi||_{L^2} = 1$. The wavepacket transform with window $\phi$ is the continuous linear mapping

$$U_\phi : \mathcal{S}(\mathbb{R}^n) \longrightarrow \widehat{S}(\mathbb{R}^{2n})$$

defined by

$$U_\phi\psi(z) = (2\pi\hbar)^{n/2}W(\psi, \phi)(z). \tag{6.15}$$

Taking into account the definition of the cross-Wigner transform, we have

$$U_\phi \psi(z) = \left(\frac{2}{\pi\hbar}\right)^{n/2} (\widehat{R}(z)\psi|\phi)_{L^2}. \tag{6.16}$$

Explicitly

$$U_\phi \psi(z) = \left(\frac{1}{2\pi\hbar}\right)^{n/2} \int_{\mathbb{R}^n} e^{-\frac{i}{\hbar}p\cdot y} \psi\left(x + \frac{1}{2}y\right) \overline{\phi\left(x - \frac{1}{2}y\right)} dy. \tag{6.17}$$

Using the properties of the cross-Wigner transform (see Section 2.1.4 in Chapter 2), the wavepacket transform extends into an automorphism

$$U_\phi : \mathcal{S}'(\mathbb{R}^n) \longrightarrow \mathcal{S}'(\mathbb{R}^{2n})$$

whose inverse $(U_\phi)^{-1}$ is calculated as follows: if $U_\phi \psi = \Psi$ then

$$\psi(x) = \frac{1}{(\gamma|\phi)_{L^2}} \left(\frac{2}{\pi\hbar}\right)^{n/2} \int_{\mathbb{R}^{2n}} \Psi(z_0)\widehat{R}(z_0)\gamma(x)dz \tag{6.18}$$

for each $\gamma \in \mathcal{S}(\mathbb{R}^n)$ such that $(\gamma|\phi)_{L^2} \neq 0$.

A well-known topic in harmonic analysis is the Bargmann transform

$$B\psi(z) = 2^{n/4} \int_{\mathbb{R}^n} e^{2\pi u\cdot z - \pi u^2 - \frac{\pi}{2}z^2} \psi(u)du. \tag{6.19}$$

It is an integral transform defined on $L^2(\mathbb{R}^n)$ and taking its values in the space of complex functions on $\mathbb{C}^n$. The Bargmann transform is that related to the Gabor transform is defined in Chapter 2

$$V_{\phi_0}\psi(x, -p) = e^{i\pi x\cdot p} B\psi(z)e^{-\pi|z|^2/2}, \tag{6.20}$$

where $z = x + ip$ and $\phi_0(z) = 2^{n/4}e^{-\pi|x|^2}$ (see [52, pp. 53–54]). From this it is easy to find the relationship between the Bargmann transform and the wavepacket transform $U_{\phi_0}$.

### 6.3.2. Properties of the wavepacket transform

As mentioned above, the wavepacket transform $U_\phi$ is a partial isometry onto a closed subspace of the Hilbert space $L^2(\mathbb{R}^{2n})$.

**Proposition 72.** *For every $\phi \in \mathcal{S}(\mathbb{R}^n)$, the mapping $U_\phi$ is an isometry of $L^2(\mathbb{R}^n)$ onto a closed subspace $\mathcal{H}_\phi$ of $L^2(\mathbb{R}^{2n})$. The adjoint operator*

$$U_\phi^* : L^2(\mathbb{R}^{2n}) \longrightarrow L^2(\mathbb{R}^n)$$

*of $U_\phi$ is given by the formula*

$$U_\phi^* \Psi(x) = \left(\frac{2}{\pi\hbar}\right)^{n/2} \int_{\mathbb{R}^{2n}} \Psi(z_0) \widehat{R}(z_0)\phi(x) dz_0. \qquad (6.21)$$

**Proof.** Let $\psi \in \mathcal{S}(\mathbb{R}^n)$. Using Moyal's identity ((6.1), Proposition 65 in Chapter 6), we have

$$(U_\phi\psi|U_\phi\psi')_{L^2(\mathbb{R}^{2n})} = (\psi|\psi')_{L^2}(\phi|\phi)_{L^2},$$

that is, since $\|\phi\|_{L^2} = 1$,

$$(U_\phi\psi|U_\phi\psi')_{L^2(\mathbb{R}^{2n})} = (\psi|\psi')_{L^2}$$

hence the operator $U_\phi$ extends into an isometry of $L^2(\mathbb{R}^n)$ onto a subspace $\mathcal{H}_\phi$ of $L^2(\mathbb{R}^{2n})$; that subspace is closed since it is the image of $L^2(\mathbb{R}^n)$. Let us prove formula (6.21) for the adjoint of $U_\phi$. By definition of the adjoint of an operator we have

$$(U_\phi^*\Psi|\psi)_{L^2} = (\Psi|U_\phi\psi)_{L^2(\mathbb{R}^{2n})}$$

hence it suffices to show that

$$\left(\frac{2}{\pi\hbar}\right)^{n/2} \int_{\mathbb{R}^{2n}} \Psi(z_0)(\widehat{R}(z_0)\phi|\psi)_{L^2} dz_0 = (\Psi|U_\phi\psi)_{L^2(\mathbb{R}^{2n})}. \qquad (6.22)$$

Now, using definitions (2.1) and (6.15) we can write

$$(\widehat{R}(z_0)\phi|\psi)_{L^2} = (\pi\hbar)^n W(\phi, \psi)(z_0) = \left(\frac{\pi\hbar}{2}\right)^{n/2} \overline{U_\phi\psi(z_0)}$$

and hence

$$\left(\frac{2}{\pi\hbar}\right)^{n/2} \int_{\mathbb{R}^{2n}} \Psi(z_0)(\widehat{R}(z_0)\phi|\psi)_{L^2} dz_0 = (\Psi|U_\phi\psi)_{L^2(\mathbb{R}^{2n})}$$

which was to be proven.                                              $\square$

**Corollary 73.** *The operator $U_\phi^*U_\phi$ is the identity on $L^2(\mathbb{R}^n)$ and*

$$\Pi_\phi = U_\phi U_\phi^* : L^2(\mathbb{R}^{2n}) \longrightarrow L^2(\mathbb{R}^{2n})$$

*is the orthogonal projection of $L^2(\mathbb{R}^{2n})$ onto the Hilbert space $\mathcal{H}_\phi$.*

**Proof.** To prove that $U_\phi^* U_\phi$ is the identity on $L^2(\mathbb{R}^n)$ we choose $\psi \in L^2(\mathbb{R}^n)$ and observe that for every $\psi' \in L^2(\mathbb{R}^n)$ we have

$$(U_\phi^* U_\phi \psi | \psi')_{L^2} = (U_\phi \psi | U_\phi \psi')_{L^2(\mathbb{R}^{2n})} = (\psi | \psi')_{L^2};$$

it follows that $U_\phi^* U_\phi \psi = \psi$. We have $\Pi_\phi = \Pi_\phi^*$ and $\Pi_\phi \Pi_\phi^* = \Pi_\phi$ hence $\Pi_\phi$ is an orthogonal projection. Since $U_\phi^* U_\phi$ is the identity on $L^2(\mathbb{R}^n)$, the range of $U_\phi^*$ is $L^2(\mathbb{R}^n)$ and that of $\Pi_\phi$ is therefore $\mathcal{H}_\phi$. □

A very important property of wavepacket transforms is that they allow to construct orthonormal bases of $L^2(\mathbb{R}^{2n})$ from orthonormal bases in $L^2(\mathbb{R}^n)$. To prove this we need a lemma.

**Lemma 74.** *Let* $\Psi \in L^2(\mathbb{R}^{2n})$. *If we have*

$$\int_{\mathbb{R}^{2n}} \Psi(z) W(\psi, \phi)(z) dz = 0$$

*for all* $\phi, \psi \in \mathcal{S}(\mathbb{R}^n)$, *then* $\Psi = 0$.

**Proof.** Recall (formula (6.21)) that the adjoint of $U_\phi$ is given by

$$U_\phi^* \Psi = \left(\frac{2}{\pi\hbar}\right)^{n/2} \int_{\mathbb{R}^{2n}} \Psi(z_0) \widehat{R}(z_0) \phi \, dz_0.$$

Let now $\psi \in \mathcal{S}(\mathbb{R}^n)$; we have, by definition of the cross-Wigner transform,

$$(U_\phi^* \Psi | \psi)_{L^2} = \left(\frac{2}{\pi\hbar}\right)^{n/2} \int_{\mathbb{R}^{2n}} \Psi(z)(\widehat{R}(z)\phi | \psi)_{L^2} dz$$

$$= (2\pi\hbar)^{n/2} \int_{\mathbb{R}^{2n}} \Psi(z) W(\psi, \phi)(z) dz.$$

Viewing $\Psi \in L^2(\mathbb{R}^{2n})$ as the Weyl symbol of an operator $\widehat{A}_\Psi$, we have, in view of formula (4.17) in Chapter 4,

$$\int_{\mathbb{R}^{2n}} \Psi(z) W(\psi, \phi)(z) dz = (\widehat{A}_\Psi \psi | \phi)_{L^2}$$

and the condition $U_\phi^* \Psi = 0$ for all $\phi \in \mathcal{S}(\mathbb{R}^n)$ is thus equivalent to $(\widehat{A}_\Psi \psi | \phi)_{L^2} = 0$ for all $\phi, \psi \in \mathcal{S}(\mathbb{R}^n)$. Hence $\widehat{A}_\Psi \psi = 0$ for all $\psi$, that is $\widehat{A}_\Psi = 0$. Since the Weyl correspondence is one-to-one we have $\Psi = 0$. □

**Proposition 75.** *Let* $(\phi_j)_j$ *and* $(\psi_j)_j$ *be orthonormal bases of* $L^2(\mathbb{R}^n)$. *Then the family* $(\Phi_{j,k})_{j,k}$ *with* $\Phi_{j,k} = U_{\phi_j} \psi_k$ *is an orthonormal basis of* $L^2(\mathbb{R}^{2n})$.

**Proof.** In view of Moyal's identity

$$(\Phi_{j,k}|\Phi_{j',k'})_{L^2(\mathbb{R}^{2n})} = (U_{\phi_j}\psi_k|U_{\phi_{j'}}\psi_{k'})_{L^2(\mathbb{R}^{2n})}$$

$$= (2\pi\hbar)^n(W(\psi_k,\phi_j)|W(\psi_{k'},\phi_{j'}))_{L^2(\mathbb{R}^{2n})}$$

$$= (\psi_k|\psi_{k'})_{L^2}\overline{(\phi_j|\phi_{j'})_{L^2}}$$

hence the $\Phi_{j,k}$ form an orthonormal system in $L^2(\mathbb{R}^{2n})$. To show that it is a basis, it is sufficient to show that if $\Psi \in L^2(\mathbb{R}^{2n})$ is orthogonal to each $\Phi_{j,k}$ then $\Psi = 0$. Assume that $(\Psi|\Phi_{jk})_{L^2(\mathbb{R}^{2n})} = 0$ for all indices $j,k$. Since we have

$$(\Psi|\Phi_{jk})_{L^2(\mathbb{R}^{2n})} = (\Psi|U_{\phi_j}\psi_k)_{L^2(\mathbb{R}^{2n})}$$

$$= (U_{\phi_j}^*\Psi|\psi_k)_{L^2(\mathbb{R}^{2n})},$$

it follows that we must have $U_{\phi_j}^*\Psi = 0$ for all $j$ since $(\psi_k)_k$ is a basis of $L^2(\mathbb{R}^n)$. Using the sesquilinearity of $U_\phi$ in $\phi$, this implies that we have $U_\phi^*\Psi = 0$ for all $\phi \in L^2(\mathbb{R}^n)$ since $(\phi_j)_j$ is a basis of $L^2(\mathbb{R}^n)$. But this implies $\Psi = 0$ in view of Lemma 74.                                      $\square$

## Main Formulas in Chapter 6

| | |
|---|---|
| Moyal CWT | $(W(\psi,\phi)|W(\psi',\phi'))_{L^2} = \left(\frac{1}{2\pi\hbar}\right)^n (\psi|\psi')_{L^2}\overline{(\phi|\phi')_{L^2}}$ |
| Moyal WT | $\|W\psi\|_{L^2} = \left(\frac{1}{2\pi\hbar}\right)^{n/2}\|\psi\|_{L^2}$ |
| Moyal CAF | $(A(\psi,\phi)|A(\psi',\phi'))_{L^2} = \left(\frac{1}{2\pi\hbar}\right)^n (\psi|\psi')_{L^2}\overline{(\phi|\phi')_{L^2}}$ |
| Reconstruction, 1 | $\psi = \frac{2^n}{(\phi|\gamma)_{L^2}} \int_{\mathbb{R}^{2n}} W(\psi,\phi)(z_0)\widehat{R}(z_0)\gamma dz_0$ |
| Reconstruction, 2 | $\psi = \frac{1}{(\phi|\gamma)_{L^2}} \int_{\mathbb{R}^{2n}} A(\psi,\phi)(z_0)\widehat{T}(z_0)\gamma dz_0$ |
| WPT | $U_\phi\psi = (2\pi\hbar)^{n/2}W(\psi,\phi)$ |
| Adjoint WPT | $U_\phi^*\Psi = \left(\frac{2}{\pi\hbar}\right)^{n/2} \int_{\mathbb{R}^{2n}} \Psi(z_0)\widehat{R}(z_0)\phi dz_0$ |
| Properties | $U_\phi^*U_\phi = I, \quad U_\phi U_\phi^* = \Pi_\phi$ |

# Chapter 7

# The Feichtinger Algebra

**Summary 76.** The Feichtinger algebra is a functional space whose elements are defined in terms of the absolute integrability of their Wigner transform. It is an algebra for both pointwise multiplication and convolution. It plays a fundamental role in functional and time-frequency analysis.

In this chapter, we deal with a function space that can be defined solely in terms of the Wigner transform, the Feichtinger algebra $S_0(\mathbb{R}^n)$. It was introduced by Hans Feichtinger [22–25] in the beginning of the 1980s; it is a particular case of the more general notion of *modulation space* we do not study here (see [52] for a detailed exposition). The Feichtinger algebra and its dual $S_0'(\mathbb{R}^n)$ are tools of choice in the study of functional properties in time-frequency analysis and the theory of pseudodifferential operators; its importance in quantum mechanics has begun to be understood more recently.

## 7.1. Definition and First Properties

The Feichtinger algebra is usually defined using the Gabor transform. Using, instead, the cross-Wigner transform has many practical advantages, especially when one wants to establish properties of metaplectic invariance.

### 7.1.1. Definition of $S_0(\mathbb{R}^n)$

We recall that the Wigner transform $W\psi$ is defined for $\psi \in L^2(\mathbb{R}^n)$.

**Definition 77.** The Feichtinger algebra $S_0(\mathbb{R}^n)$ (also denoted $M^1(\mathbb{R}^n)$) consists of all functions $\psi \in L^2(\mathbb{R}^n)$ such that $W\psi \in L^1(\mathbb{R}^{2n})$:

$$||W\psi||_{L^1(\mathbb{R}^{2n})} = \int_{\mathbb{R}^{2n}} |W\psi(z)|dz < \infty.$$

Here is an elementary example which shows that the elements of $S_0(\mathbb{R}^n)$ need not be differentiable.

**Example 78.** Let $\psi \in C^0(\mathbb{R}^n)$ be defined by

$$\psi(x) = \begin{cases} 1 - |x| & \text{if } |x| \leq 1, \\ 0 & \text{if } |x| > 1. \end{cases}$$

We have $\psi \in S_0(\mathbb{R}^n)$.

We will see that $S_0(\mathbb{R}^n)$ is a *vector space*; notice that this is not at all obvious from the definition above: while it is clear that if $\psi \in S_0(\mathbb{R}^n)$ then $\lambda\psi \in S_0(\mathbb{R}^n)$ for all $\lambda \in \mathbb{C}$ (because $W(\lambda\psi) = |\lambda|^2 W\psi$), and we have to check that if $\psi, \phi \in S_0(\mathbb{R}^n)$ then $\psi + \phi \in S_0(\mathbb{R}^n)$. But the Wigner transform is not linear, in fact

$$W(\psi + \phi) = W\psi + W\phi + 2ReW(\psi, \phi).$$

It follows that we will have $\psi + \phi \in S_0(\mathbb{R}^n)$ provided that $ReW(\psi, \phi) \in L^1(\mathbb{R}^{2n})$. To prove that this is indeed the case we need following result.

**Proposition 79.** *Let* $\psi, \phi \in L^2(\mathbb{R}^n)$. *If* $W\psi \in L^1(\mathbb{R}^{2n})$ *and* $W\phi \in L^1(\mathbb{R}^{2n})$, *then* $W(\psi, \phi) \in L^1(\mathbb{R}^{2n})$. *There exists a constant* $C > 0$ *such that*

$$||W\psi||_{L^1(\mathbb{R}^{2n})} \leq C||W(\psi, \phi)||_{L^1(\mathbb{R}^{2n})}. \tag{7.1}$$

**Proof.** Recall that we proved in Chapter 6 (Proposition 69) that if $\psi, \theta \in L^2(\mathbb{R}^n)$, $\theta \neq 0$, then the integral

$$\int_{\mathbb{R}^{2n}} W(\psi, \theta)(z_0)\widehat{R}(z_0)\gamma(x)dz_0$$

defines a function in $L^2(\mathbb{R}^n)$, and we have

$$\psi(x) = \frac{2^n}{(\theta|\gamma)_{L^2}} \int_{\mathbb{R}^{2n}} W(\psi, \theta)(z_0)\widehat{R}(z_0)\gamma(x)dz_0$$

for every $\gamma \in L^2(\mathbb{R}^n)$ such that $(\theta|\gamma)_{L^2} \neq 0$. Applying the Grossmann–Royer operator $\widehat{R}(z_1)$ to both sides of this equality and using formula (1.16) in Chapter 1 we get

$$\widehat{R}(z_1)\psi(x) = \frac{2^n}{(\theta|\gamma)_{L^2}} \int_{\mathbb{R}^{2n}} W(\psi, \theta)(z_0)\widehat{R}(z_1)\widehat{R}(z_0)\gamma(x)dz_0$$

$$= \frac{2^n}{(\theta|\gamma)_{L^2}} \int_{\mathbb{R}^{2n}} W(\psi, \theta)(z_0)e^{\frac{2i}{\hbar}\sigma(z_0, z_1)}\widehat{T}(2(z_1 - z_0))\gamma(x)dz_0.$$

Let now $\phi \in L^2(\mathbb{R}^n)$; multiplying both sides of the equality above by $\overline{\phi(x)}$ and integrating with respect to $x$ we get

$$(\widehat{R}(z_1)\psi|\phi)_{L^2} = \frac{2^n}{(\theta|\gamma)_{L^2}} \int_{\mathbb{R}^{2n}} e^{\frac{2i}{\hbar}\sigma(z_0, z_1)}W(\psi, \theta)(z_0)$$

$$\times (\widehat{T}(2(z_1 - z_0))\gamma|\phi)_{L^2}dz_0. \tag{7.2}$$

We next observe that by definition of the cross-Wigner transform we have

$$(\widehat{R}(z_1)\psi|\phi)_{L^2} = (\pi\hbar)^n W(\psi, \phi)(z_1)$$

and, similarly, by definition of the cross-ambiguity function, and using the identity (3.17) in Chapter 3,

$$(\widehat{T}(2(z_1 - z_0))\gamma|\phi)_{L^2} = (2\pi\hbar)^n A(\gamma, \phi)(2(z_1 - z_0))$$

$$= (\pi\hbar)^n W(\gamma, \phi^{\vee})(z_1 - z_0),$$

where $\phi^{\vee}(x) = \phi(-x)$. It follows that we can rewrite the equality (7.2) as

$$W(\psi, \phi)(z_1) = \frac{2^n}{(\theta|\gamma)_{L^2}} \int_{\mathbb{R}^{2n}} e^{\frac{2i}{\hbar}\sigma(z_0, z_1)}W(\psi, \theta)(z_0)W(\gamma, \phi^{\vee})(z_1 - z_0)dz_0$$

and hence

$$|W(\psi, \phi)(z_1)| \leq \frac{2^n}{|(\theta|\gamma)_{L^2}|} \int_{\mathbb{R}^{2n}} |W(\psi, \theta)(z_0)W(\gamma, \phi^{\vee})(z_1 - z_0)|dz_0.$$

Rewriting this inequality as

$$|W(\psi, \phi)|(z) \leq \frac{2^n}{|(\theta|\gamma)_{L^2}|}|W(\psi, \theta) * W(\gamma, \phi^{\vee})(z)|$$

and integrating both sides with respect to $z$ yields

$$||W(\psi, \phi)||_{L^1(\mathbb{R}^{2n})} \leq \frac{2^n}{|(\theta|\gamma)_{L^2}|} ||W(\psi, \theta) * W(\gamma, \phi^\vee)||_{L^1(\mathbb{R}^{2n})},$$

that is, taking the properties of the convolution product in consideration,

$$||W(\psi, \phi)||_{L^1(\mathbb{R}^{2n})} \leq \frac{2^n}{|(\theta|\gamma)_{L^2}|} ||W(\psi, \theta)||_{L^1(\mathbb{R}^{2n})|| ||W(\gamma, \phi^\vee)||_{L^1(\mathbb{R}^{2n})}}$$

$$(7.3)$$

hence the inequality (7.1) choosing $\theta = \psi$ and

$$C = \frac{2^n}{|(\phi|\gamma)_{L^2}|} ||W(\gamma, \theta^\vee)||_{L^1(\mathbb{R}^{2n})}.$$

$\square$

An immediate consequence of this result is the following general condition for the integrability of the cross-Wigner transform.

**Corollary 80.** *Let $\psi, \phi \in L^2(\mathbb{R}^n)$. We have $W(\psi, \phi) \in L^1(\mathbb{R}^{2n})$ if and only if both $\psi \in S_0(\mathbb{R}^n)$ and $\phi \in S_0(\mathbb{R}^n)$.*

**Proof.** The sufficiency of the condition is just a restatement of Proposition 79. It follows from the inequality (7.1) in the same proposition that we have both

$$||W\psi||_{L^1(\mathbb{R}^{2n})} \leq C||W(\psi, \phi)||_{L^1(\mathbb{R}^{2n})},$$

$$||W\phi||_{L^1(\mathbb{R}^{2n})} \leq C'||W(\psi, \phi)||_{L^1(\mathbb{R}^{2n})}$$

for some constants $C$ and $C'$. The necessity follows.          $\square$

Another immediate consequence of the above results is as follows.

**Corollary 81.** *Let $\psi \in L^2(\mathbb{R}^n)$.*

(i) *We have $\psi \in S_0(\mathbb{R}^n)$ if and only if there exists $\phi \in S_0(\mathbb{R}^n)$ such that $W(\psi, \phi) \in L^1(\mathbb{R}^{2n})$.*
(ii) *If $\psi \in S_0(\mathbb{R}^n)$ then $W(\psi, \phi) \in L^1(\mathbb{R}^{2n})$ for all $\phi \in S_0(\mathbb{R}^n)$.*

An important property, which is immediately visible using the definition of $S_0(\mathbb{R}^n)$ in terms of the Wigner transform, is its invariance under the action of the metaplectic group and phase space translations.

**Proposition 82.** *Let $\psi \in S_0(\mathbb{R}^n)$.*

(i) *For every $\widehat{S} \in \mathrm{Mp}(n)$ we have $\widehat{S}\psi \in S_0(\mathbb{R}^n)$.*
(ii) *For every $z_0 \in \mathbb{R}^{2n}$ we have $\widehat{T}(z_0)\psi \in S_0(\mathbb{R}^n)$.*

**Proof.** (i) By definition of $S_0(\mathbb{R}^n)$ we have $\psi \in L^2(\mathbb{R}^n)$ and $W\psi \in L^1(\mathbb{R}^n)$. Since $\widehat{S} \in \mathrm{Mp}(n)$ is unitary on $L^2(\mathbb{R}^n)$ we have $\widehat{S}\psi \in L^2(\mathbb{R}^n)$. Now, by the symplectic covariance property of the Wigner transform (formula (5.11) in Chapter 5) we have

$$W(\widehat{S}\psi)(z) = W\psi(S^{-1}z)$$

where $S \in \mathrm{Sp}(n)$ is the projection of $\widehat{S}$; since

$$\int_{\mathbb{R}^{2n}} |W\psi(S^{-1}z)|dz = \int_{\mathbb{R}^{2n}} |W\psi(z)|dz$$

(because $\det S = 1$) we thus have $W(\widehat{S}\psi) \in L^1(\mathbb{R}^n)$ and hence $\widehat{S}\psi \in S_0(\mathbb{R}^n)$ as claimed.

(ii) We have

$$W(\widehat{T}(z_0)\psi)(z) = W\psi(z - z_0)$$

(formula (2.30) in Chapter 2) hence $W(\widehat{T}(z_0)\psi) \in L^1(\mathbb{R}^{2n})$ if and only if $W\psi \in L^1(\mathbb{R}^{2n})$. $\qquad\square$

**Example 83.** The Feichtinger algebra is invariant under the Fourier transform: we have $\psi \in S_0(\mathbb{R}^n)$ if and only if $F\psi \in S_0(\mathbb{R}^n)$ since the Fourier transform is the operator $\widehat{J} \in \mathrm{Mp}(n)$ up to a unimodular constant factor. It is also invariant under linear changes of variables $x \longmapsto Lx$ (because $\widehat{M}_{L,m} \in \mathrm{Mp}(n)$).

Notice that it follows from the example above that if $\psi \in S_0(\mathbb{R}^n)$ then $\lim_{z \to 0} \psi = 0$. In fact, since $S_0(\mathbb{R}^n)$ is invariant by Fourier transform, we have $F^{-1}\psi \in S_0(\mathbb{R}^n)$; now $S_0(\mathbb{R}^n) \subset L^1(\mathbb{R}^n)$ hence $\psi = F(F^{-1}\psi)$ has limit $0$ at infinity in view of Riemann–Lebesgue's lemma.

Proposition 82 can be summarized by saying that $S_0(\mathbb{R}^n)$ is invariant under the action of the inhomogeneous metaplectic group. One proves (we will not do it here: see [39, 52]) that $S_0(\mathbb{R}^n)$ is the smallest Banach algebra containing $\mathcal{S}(\mathbb{R}^n)$ having this property.

### 7.1.2. Analytical properties of $S_0(\mathbb{R}^n)$

We have established above that the Feichtinger algebra $S_0(\mathbb{R}^n)$ indeed is a vector space. It turns out that we can equip $S_0(\mathbb{R}^n)$ with a norm (in fact, infinitely many equivalent norms); in addition $S_0(\mathbb{R}^n)$ is complete for the topology thus defined.

**Proposition 84.** *Let $\phi \in \mathcal{S}(\mathbb{R}^n)$, $\phi \neq 0$.*

(i) *The mapping $\psi \longmapsto ||\psi||_\phi$ defined by*

$$||\psi||_\phi = ||W(\psi, \phi)||_{L^1(\mathbb{R}^{2n})}$$

*is a norm on $S_0(\mathbb{R}^n)$.*

(ii) *For every $\phi' \in \mathcal{S}(\mathbb{R}^n)$, $\phi' \neq 0$ there exist constants $A, B > 0$ such that*

$$A||\psi||_\phi \leq ||\psi||_{\phi'} \leq B||\psi||_\phi$$

*for all $\psi \in S_0(\mathbb{R}^n)$; all the norms $||\cdot||_\phi$ on $S_0(\mathbb{R}^n)$ are thus equivalent.*

(iii) *The Schwartz space $\mathcal{S}(\mathbb{R}^n)$ is dense in $S_0(\mathbb{R}^n)$.*

**Proof.** (i) That $\psi \longmapsto ||\psi||_\phi$ satisfies the property $||\lambda\psi||_\phi = |\lambda| \, ||\psi||_\phi$ and the triangle inequality is obvious. Assume that $||\psi||_\phi = 0$. Then in view of the inequality (7.1) we have $W\psi = 0$ hence $\psi = 0$.

(ii) In Proposition 79, we established that given $\phi$ and $\theta$ there exists a constant $C > 0$ such that

$$||W(\psi, \phi)||_{L^1(\mathbb{R}^{2n})} \leq C||W(\psi, \theta)||_{L^1(\mathbb{R}^{2n})},$$

for all $\psi \in S_0(\mathbb{R}^n)$. This is equivalent to $||\psi||_\phi \leq C||\psi||_\theta$. Similarly, interchanging $\phi$ and $\theta$ there exists a constant $C'$ such that

$$||W(\psi, \theta)||_{L^1(\mathbb{R}^{2n})} \leq C'||W(\psi, \phi)||_{L^1(\mathbb{R}^{2n})},$$

that is $||\psi||_\theta \leq C'||\psi||_\phi$. The existence of $A$ and $B$ follows.

(iii) That $\mathcal{S}(\mathbb{R}^n) \subset S_0(\mathbb{R}^n)$ is clear since $\psi \in \mathcal{S}(\mathbb{R}^n)$ implies that $W\psi \in \mathcal{S}(\mathbb{R}^{2n}) \subset L^1(\mathbb{R}^{2n})$. Let us show that for every $\psi \in S_0(\mathbb{R}^n)$ there exists a sequence $(\psi_j)_j$ in $\mathcal{S}(\mathbb{R}^n)$ such that

$$\lim_{j \to \infty} ||\psi - \psi_j||_\phi = 0$$

for $\phi \in \mathcal{S}(\mathbb{R}^n)$; this will prove the density of $\mathcal{S}(\mathbb{R}^n)$ in $S_0(\mathbb{R}^n)$. Let $B_j = B^{2n}(j)$ be the ball in $\mathbb{R}^{2n}$ centered at the origin with radius $j$ and define for each $j$ a function $\psi_j$ by

$$\psi_j(x) = \frac{2^n}{(\gamma|\phi)_{L^2}} \int_{B_j} W(\psi,\phi)(z_0)\widehat{R}(z_0)\gamma(x)dz_0, \qquad (7.4)$$

where $\gamma \in L^2(\mathbb{R}^n)$ is chosen such that $(\gamma|\phi)_{L^2} \neq 0$. Each $\psi_j$ belongs to $\mathcal{S}(\mathbb{R}^n)$: for all multi-indices $\alpha, \beta \in \mathbb{N}^n$ we have

$$x^\alpha \partial_x^\beta \psi_j(x) = \frac{2^n}{(\gamma|\phi)_{L^2}} \int_{B_j} W(\psi,\phi)(z_0)\widehat{R}(z_0)(x^\alpha \partial_x^\beta \gamma)(x)dz_0$$

so that, taking the suprema in $x$,

$$\sup_{\mathbb{R}^n} |x^\alpha \partial_x^\beta \psi_j| \leq \frac{2^n}{(\gamma|\phi)_{L^2}} \int_{B_j} |W(\psi,\phi)(z_0)|dz_0 \left( \sup_{\mathbb{R}^n} |x^\alpha \partial_x^\beta \gamma| \right).$$

Since $\psi \in S_0(\mathbb{R}^n)$ we have

$$\int_{B_j} |W(\psi,\phi)(z_0)|dz_0 < \int_{\mathbb{R}^{2n}} |W(\psi,\phi)(z_0)|dz_0 < \infty$$

hence there exists $C_j > 0$ such that

$$\sup_{\mathbb{R}^n} |x^\alpha \partial_x^\beta \psi_j| \leq C_j \sup_{\mathbb{R}^n} |x^\alpha \partial_x^\beta \gamma| < \infty,$$

which shows that we have $\psi_j \in \mathcal{S}(\mathbb{R}^n)$. Let us show that $\lim_{j\to\infty} \psi_j = \psi$ in $S_0(\mathbb{R}^n)$. Recalling (Chapter 6, Section 6.3) that the adjoint of the wavepacket transform

$$U_\phi \psi(z) = (2\pi\hbar)^{n/2} W(\psi,\phi)(z)$$

is given by

$$U_\phi^* \Psi(x) = \left(\frac{2}{\pi\hbar}\right)^{n/2} \int_{\mathbb{R}^{2n}} \Psi(z_0)\widehat{R}(z_0)\phi(x)dz_0,$$

we can rewrite definition (7.4) of $\psi_j$ as

$$\psi_j = \frac{2^n}{(\gamma|\phi)_{L^2}} \int_{B_j} \Psi_j(z_0)\widehat{R}(z_0)\gamma dz_0 = U_\phi^* \Psi_j,$$

where $\Psi_j = (U_\phi \psi) \chi_j$ and $\chi_j$ is the characteristic function of the ball $B_j$. Since we have $U_\phi^* U_\phi = I_{\mathrm{d}}$ on $L^2(\mathbb{R}^n)$, it follows that

$$||\psi - \psi_j||_\phi = ||U_\phi^*(U_\phi \psi - \Psi_j)||_\phi \leq C||U_\phi \psi - \Psi_j||_{L^1(\mathbb{R}^{2n})}.$$

Noting that we have $U_\phi \psi - \Psi_j = U_\phi \psi (1 - \chi_j)$ we get

$$||U_\phi \psi - \Psi_j||_{L^1(\mathbb{R}^{2n})} = ||U_\phi \psi (1 - \chi_j)||_{L^1(\mathbb{R}^{2n})}$$

$$= (2\pi\hbar)^{n/2} \int_{\mathbb{R}^{2n}} |W(\psi, \phi)(z)|(1 - \chi_j(z)) dz$$

and hence $\lim_{j \to \infty} ||U_\phi \psi - \Psi_j||_\phi = 0$ by the dominated convergence theorem. $\qquad\square$

Let us next prove the completeness of the normed vector space $S_0(\mathbb{R}^n)$.

**Proposition 85.** *The normed space $S_0(\mathbb{R}^n)$ is complete, hence a Banach space.*

**Proof.** (Cf. [52, Theorem 11.3.5].) Let $(\psi_j)$ be a Cauchy sequence in $S_0(\mathbb{R}^n)$; setting $\Psi_j = W(\psi_j, \phi)$ we get a Cauchy sequence in $L^1(\mathbb{R}^{2n})$. The latter being a Banach space, there exists $\Psi \in L^1(\mathbb{R}^{2n})$ such that

$$\lim_{j \to \infty} ||\Psi - W(\psi_j, \phi)||_{L^1(\mathbb{R}^{2n})} = 0.$$

Let us define a function $\psi$ by the formula

$$\psi(x) = \frac{2^n}{||\phi||_{L^2}} \int_{\mathbb{R}^{2n}} \Psi(z_0) \widehat{R}(z_0) \phi(x) dz_0 \qquad (7.5)$$

as in the proof of Proposition 84(iii), one then shows that $\psi \in S_0(\mathbb{R}^n)$ and that

$$\lim_{j \to \infty} ||\psi - \psi_j||_\phi = \lim_{j \to \infty} ||W(\psi - \psi_j, \phi)||_{L^1(\mathbb{R}^{2n})}$$

$$= \lim_{j \to \infty} ||\Psi - W(\psi_j, \phi)||_{L^1(\mathbb{R}^{2n})}$$

$$= 0$$

hence the Cauchy sequence $(\psi_j)$ converges to $\psi \in S_0(\mathbb{R}^n)$. $\qquad\square$

### 7.1.3. The algebra property of $S_0(\mathbb{R}^n)$

We are going to show that the Feichtinger algebra is a Banach algebra for both convolution and pointwise multiplication. We begin by considering the convolution of two elements of $S_0(\mathbb{R}^n)$.

**Proposition 86.** *Suppose that $\psi \in L^1(\mathbb{R}^n)$ and $\psi' \in S_0(\mathbb{R}^n)$. Then $\psi * \psi' \in S_0(\mathbb{R}^n)$ and we have*

$$||\psi * \psi'||_\phi \leq ||\psi||_{L^1} ||\psi'||_\phi \tag{7.6}$$

*for every $\phi \in \mathcal{S}(\mathbb{R}^n)$. Thus, if $\psi \in L^1(\mathbb{R}^n)$ and $\psi' \in S_0$ then $\psi * \psi' \in S_0(\mathbb{R}^n)$.*

**Proof.** The definition of the cross-Wigner transform

$$W(\psi, \phi)(z) = \left(\frac{1}{\pi\hbar}\right)^n (\widehat{R}(z)\psi|\phi)_{L^2}$$

can be rewritten

$$W(\psi, \phi)(z_0) = \left(\frac{1}{\pi\hbar}\right)^n e^{-\frac{2i}{\hbar}p_0 \cdot x_0} \int_{\mathbb{R}^n} \psi(2x_0 - x)\phi_{p_0}(x)dx$$

with $\phi_{p_0}(x) = e^{\frac{2i}{\hbar}p_0 \cdot x}\phi(x)$, that is

$$W(\psi, \phi)(z_0) = \left(\frac{1}{\pi\hbar}\right)^n e^{-\frac{2i}{\hbar}p_0 \cdot x_0} \psi * \phi_{p_0}(2x_0). \tag{7.7}$$

It follows that

$$||\psi||_\phi = \left(\frac{1}{2\pi\hbar}\right)^n \int_{\mathbb{R}^n} ||\psi * \phi_{p_0}||_{L^1} dp_0. \tag{7.8}$$

Formula (7.7) now shows that

$$W(\psi * \psi', \phi)(z_0) = \left(\frac{1}{\pi\hbar}\right)^n e^{-\frac{2i}{\hbar}p_0 \cdot x_0} \psi * \psi' * \phi_{p_0}(2x_0)$$

and hence, by (7.8),

$$||\psi * \psi'||_\phi = \left(\frac{1}{2\pi\hbar}\right)^n \int_{\mathbb{R}^n} ||\psi * (\psi' * \phi_{p_0})||_{L^1} dp_0.$$

Since $L^1(\mathbb{R}^n)$ is a convolution algebra, we have

$$||\psi * (\psi' * \phi_{p_0})||_{L^1} \leq ||\psi||_{L^1} ||\psi' * \phi_{p_0}||_{L^1},$$

and we obtain the inequality

$$||\psi * \psi'||_\phi \le \left(\frac{1}{2\pi\hbar}\right)^n ||\psi||_{L^1} \int_{\mathbb{R}^n} ||\psi' * \phi_{p_0}||_{L^1} dp_0$$

that is, using again (7.8),

$$||\psi * \psi'||_\phi \le ||\psi||_{L^1} ||\psi||_\phi$$

which we set out to prove. $\square$

**Corollary 87.** *The Banach space $S_0(\mathbb{R}^n)$ is an algebra for both pointwise multiplication: if $\psi$ and $\psi'$ are in $S_0(\mathbb{R}^n)$ then $\psi\psi' \in S_0(\mathbb{R}^n)$.*

**Proof.** Since $\psi\psi'$ and $\psi * \psi'$ are interchangeable by the Fourier transform $F$, and $S_0(\mathbb{R}^n)$ being invariant under Fourier transforms (see Example 83) it is sufficient to show that $\psi * \psi' \in S_0(\mathbb{R}^n)$ if $\psi \in S_0(\mathbb{R}^n)$ and $\psi' \in S_0(\mathbb{R}^n)$. But this follows from the inequality (7.6) since $S_0(\mathbb{R}^n) \subset L^1(\mathbb{R}^n)$. $\square$

## 7.2. The Dual Space $S_0'(\mathbb{R}^n)$

### 7.2.1. Description of $S_0'(\mathbb{R}^n)$

Let us denote by $S_0'(\mathbb{R}^n)$ the dual Banach space of $S_0(\mathbb{R}^n)$; by definition it is the space of all bounded linear functionals on $S_0(\mathbb{R}^n)$.

The following result characterizes $S_0'(\mathbb{R}^n)$.

**Proposition 88.** *The dual $S_0'(\mathbb{R}^n)$ of the Feichtinger algebra consists of all $\psi \in S'(\mathbb{R}^n)$ such that $W(\psi,\phi) \in L^\infty(\mathbb{R}^{2n})$ for one (and hence all) $\phi \in S_0(\mathbb{R}^n)$; the duality bracket is given by the pairing*

$$(\psi, \psi') = \int_{\mathbb{R}^{2n}} W(\psi,\phi)(z)\overline{W(\psi',\phi)(z)}dz \qquad (7.9)$$

*and the formula*

$$||\psi||_{\phi,S_0'} = \sup_{z\in\mathbb{R}^{2n}} |W(\psi,\phi)(z)| \qquad (7.10)$$

*defines a norm on the Banach space $S_0'(\mathbb{R}^n)$.*

**Proof.** It is based on the fact that $L^\infty(\mathbb{R}^{2n})$ is the dual space of $L^1(\mathbb{R}^{2n})$; see [52, §11.3]. $\square$

It readily follows from this characterization that:

**Proposition 89.** *The Dirac distribution $\delta$ is in $S_0'(\mathbb{R}^n)$; more generally $\delta_a \in S_0'(\mathbb{R}^n)$ where $\delta_a(x) = \delta(x-a)$.*

**Proof.** We have

$$W(\delta, \phi)(z_0) = \left(\frac{1}{\pi\hbar}\right)^n \langle \widehat{T}_{\mathrm{GR}}(z_0)\delta, \overline{\phi} \rangle$$

(formula (2.4)) and

$$\widehat{R}(z_0)\delta(x) = e^{\frac{2i}{\hbar}p_0 \cdot (x - x_0)}\delta(2x_0 - x) = e^{\frac{2i}{\hbar}p_0 \cdot x_0}\delta(2x_0 - x).$$

It follows that

$$W(\delta, \phi)(z_0) = \left(\frac{1}{\pi\hbar}\right)^n e^{\frac{2i}{\hbar}p_0 \cdot x_0}\overline{\phi}(2x_0)$$

and hence

$$|W(\delta, \phi)(z_0)| \leq \left(\frac{1}{\pi\hbar}\right)^n ||\phi||_\infty.$$

It follows that $\delta \in S_0'(\mathbb{R}^n)$. That we also have $\delta_a \in S_0'(\mathbb{R}^n)$ is proven by a similar argument (alternatively one can use formula (2.29) in Proposition 22). □

**Example 90.** More generally, the "Dirac comb" $\sum_{a \in \mathbb{Z}^n} \delta_a$ belongs to $S_0'(\mathbb{R}^n)$.

### 7.2.2. The Gelfand triple $(S_0, L^2, S_0')$

The prototypical example of a Gelfand triple is

$$(\mathcal{S}(\mathbb{R}^n), L^2(\mathbb{R}^n), \mathcal{S}'(\mathbb{R}^n)).$$

Another example is easily obtained using the Feichtinger algebra and its dual:

$$(S_0(\mathbb{R}^n), L^2(\mathbb{R}^n), S_0'(\mathbb{R}^n))$$

is a Banach Gelfand triple.

Let us begin by defining rigorously the notion of Banach Gelfand triple.

**Definition 91.** A Banach Gelfand triple $(\mathcal{B}, \mathcal{H}, \mathcal{B}')$ consists of a Banach space $\mathcal{B}$ which is continuously and densely embedded into a Hilbert space $\mathcal{H}$, which in turn is $w^*$-continuously and densely embedded into the dual Banach space $\mathcal{B}'$.

In this definition one identifies $\mathcal{H}$ with its dual $\mathcal{H}^*$ and the scalar product on $\mathcal{H}$ thus extends in a natural way into a pairing between $\mathcal{B} \subset \mathcal{H}$ and $\mathcal{B}' \supset \mathcal{H}$.

The use of the Gelfand triple $(S_0(\mathbb{R}^n), L^2(\mathbb{R}^n), S_0'(\mathbb{R}^n))$ not only offers a better description of self-adjoint operators but it also allows a simplification of many proofs. Here is a typical situation. Given a Gelfand triple $(\mathcal{B}, \mathcal{H}, \mathcal{B}')$ one proves that every self-adjoint operator $A : \mathcal{B} \longrightarrow \mathcal{B}$ has a complete family of generalized eigenvalues $(\psi_\alpha)_\alpha = \{\psi_\alpha \in \mathcal{B}' : \alpha \in \mathbb{A}\}$ ($\mathbb{A}$ an index set), defined as follows: for every $\alpha \in \mathbb{A}$ there exists $\lambda_\alpha \in \mathbb{C}$ such that

$$(\psi_\alpha, A\phi) = \lambda_\alpha(\psi_\alpha, \phi) \quad \text{for every } \phi \in \mathcal{B}.$$

Completeness of the family $(\psi_\alpha)_\alpha$ means that there exists at least one $\psi_\alpha$ such that $(\psi_\alpha, \phi) \neq 0$ for every $\phi \in \mathcal{B}$. A basic example, in the case $n = 1$, is the operator $\widehat{x}$ of multiplication by $x$. This operator has no eigenfunctions in $L^2(\mathbb{R})$ (it is unbounded), but since $\widehat{x}\delta_a(x) = x\delta(x-a) = a\delta_a(x)$ every $a \in \mathbb{R}$ is a generalized eigenvalue (with associated eigenfunction $\delta_a \in S_0'(\mathbb{R})$).

Given a Gelfand triple $(\mathcal{B}, \mathcal{H}, \mathcal{B}')$ every $\phi \in \mathcal{B}$ has an expansion with respect to the generalized eigenvectors $\psi_\alpha$ which generalizes the usual expansion with respect to a basis of eigenvectors. A classic example is the following: consider the Gelfand triple $(S_0(\mathbb{R}^n), L^2(\mathbb{R}^n), S_0'(\mathbb{R}^n))$ and choose $\widehat{A} = -i\hbar\partial_{x_j}$. The generalized eigenvalues of $\widehat{A}$ are the functions $\chi_p(x) = e^{ip \cdot x/\hbar}$ ($p \in \mathbb{R}^n$) and the corresponding expansion can be written as the Fourier inversion formula

$$\psi(x) = \left(\frac{1}{2\pi\hbar}\right)^{n/2} \int_{\mathbb{R}^n} e^{\frac{i}{\hbar}p \cdot x} F\psi(p)dp. \tag{7.11}$$

## Main Formulas in Chapter 7

| | |
|---|---|
| $\psi \in S_0(\mathbb{R}^n)$ | if $\psi \in L^2(\mathbb{R}^n)$ and $||W\psi||_{L^1(\mathbb{R}^{2n})} < \infty$ |
| $\psi \in L^2(\mathbb{R}^n)$, $\psi \in S_0(\mathbb{R}^n)$ if and only if | $\exists\phi \in S_0(\mathbb{R}^n)$: $W(\psi, \phi) \in L^1(\mathbb{R}^{2n})$ |
| $\psi \in S_0(\mathbb{R}^n)$ | $\widehat{S}\psi \in S_0(\mathbb{R}^n), \forall\widehat{S} \in \text{Mp}(n)$ |
| $\psi \in S_0(\mathbb{R}^n)$ | $\widehat{T}(z_0)\psi \in S_0(\mathbb{R}^n)$ |
| Norm on $S_0(\mathbb{R}^n)$ | $||\psi||_\phi = ||W(\psi, \phi)||_{L^1(\mathbb{R}^{2n})}$, $\phi \in \mathcal{S}(\mathbb{R}^n)$, $\phi \neq 0$ |
| Convolution algebra | $||\psi * \psi'||_\phi \leq ||\psi||_{L^1}||\psi'||_\phi$ |
| Product algebra | $\psi, \psi' \in S_0(\mathbb{R}^n) \Longrightarrow \psi\psi' \in S_0(\mathbb{R}^n)$ |
| Duality of $S_0'(\mathbb{R}^n)$ and $S_0(\mathbb{R}^n)$ | $(\psi, \psi') = \int_{\mathbb{R}^{2n}} W(\psi, \phi)(z)\overline{W(\psi', \phi)(z)}dz$ |
| Norm of $S_0'(\mathbb{R}^n)$ | $||\psi||_{\phi, S_0'} = \sup_{z \in \mathbb{R}^{2n}} |W(\psi, \phi)(z)|$ |

# Chapter 8

# The Cohen Class

**Summary 92.** The Wigner transform is the simplest example of a quasi-distribution. The Husimi distribution is another well-known example. It is possible to generate infinitely many quasi-distributions having similar properties; they belong to the so-called Cohen class. One of the most important elements of this class is the Born–Jordan distribution.

The lack of positivity of the Wigner transform $W\psi$ has led to a search for alternative transforms $Q\psi$; another reason which has prompted this quest is the lack of linearity of the Wigner transform, which leads to the appearance of the cross-terms $W(\psi_j, \psi_k)$ when calculates the Wigner transform of a linear superposition $\sum_j \lambda_j \psi_j$ (see formula (2.19) in Chapter 2). While the interpretation of these cross-terms is relatively straightforward in quantum mechanics as "sub-Planck effects", they are a nuisance for specialists in time-frequency analysis and signal theory (to witness the impressive ever-growing list of publications continuing to appear on the topic of "interference elimination").

## 8.1. Definition

We are following here de Gosson [41] with some modifications and amendments.

We begin by giving a quite general definition (cf. [52]) of the Cohen class $CC(\mathbb{R}^n)$:

**Definition 93.** Let $Q : \mathcal{S}(\mathbb{R}^n) \times \mathcal{S}(\mathbb{R}^n) \longrightarrow \mathcal{S}'(\mathbb{R}^{2n})$ be a sesquilinear form. We say that $Q$ belongs to the Cohen class if there exists a tempered distribution $\theta \in \mathcal{S}'(\mathbb{R}^{2n})$ (the "Cohen kernel") such that

$$Q(\psi, \phi) = W(\psi, \phi) * \theta \tag{8.1}$$

for all $\psi, \phi \in \mathcal{S}(\mathbb{R}^n)$; equivalently

$$F_\sigma Q(\psi, \phi) = (2\pi\hbar)^n A(\psi, \phi) F_\sigma \theta \tag{8.2}$$

($F_\sigma$ the symplectic Fourier transform). When $\psi = \phi$ one writes $Q\psi = Q(\psi, \psi)$.

Choosing $\theta = \delta$ we have $Q(\psi, \phi) = W(\psi, \phi)$ hence the cross-Wigner transform belongs to the Cohen class. Here is a less trivial example.

**Example 94.** Choose for $\theta$ the phase space Gaussian:

$$\theta(z) = (\pi\hbar)^{-n} e^{-\frac{1}{\hbar}|z|^2}.$$

The corresponding element of the Cohen class is called the Husimi distribution.

The Husimi distribution is a well-studied member of the Cohen class; we will return to it in a moment.

The following result gives a sufficient condition for a sesquilinear form $Q$ to belong to Cohen's class.

**Proposition 95.** *Let* $Q : \mathcal{S}(\mathbb{R}^n) \times \mathcal{S}(\mathbb{R}^n) \longrightarrow \mathcal{S}(\mathbb{R}^{2n})$ *be a sesquilinear form such that*

$$Q\psi(z - z_0) = Q(\widehat{T}(z_0)\psi)(z), \tag{8.3}$$

$$|Q(\psi, \phi)(0,0)| \leq ||\psi|| \, ||\phi||, \tag{8.4}$$

*for all* $\psi, \phi$ *in* $\mathcal{S}(\mathbb{R}^n)$. *Then* $Q$ *belongs to the Cohen class.*

**Proof.** (We are following [39, 52] with some minor modifications.) The condition (8.4) means that the sesquilinear form $Q$ is bounded. It follows, using Riesz's representation theorem, that there exists a bounded operator $\widehat{A}$ on $L^2(\mathbb{R}^n)$ such that

$$Q(\psi, \phi)(0, 0) = (\widehat{A}\psi|\phi)_{L^2}.$$

Using the translation property (8.3) we then have

$$Q\psi(z_0) = Q(\widehat{T}(-z_0)\psi)(0) = \langle \widehat{A}\widehat{T}(-z_0)\psi, \widehat{T}(-z_0)\psi \rangle.$$

In view of Schwartz's kernel theorem, there exists a distribution $K \in \mathcal{S}'(\mathbb{R}^n \times \mathbb{R}^n)$ such that

$$(\phi|\widehat{A}\psi)_{L^2} = \langle K, \psi \otimes \overline{\phi} \rangle$$

for all $\psi, \phi \in \mathcal{S}(\mathbb{R}^n)$. We thus have

$$Q\psi(z_0) = \langle K, \widehat{T}(-z_0)\psi \otimes \overline{\widehat{T}(-z_0)\psi} \rangle$$

$$= \int_{\mathbb{R}^{2n}} K(x,y)\widehat{T}(-z_0)\psi(x)\overline{\widehat{T}(-z_0)\psi(y)}dxdy.$$

By definition of $\widehat{T}(z_0)$ we have

$$\widehat{T}(-z_0)\psi(x) = e^{\frac{i}{\hbar}(-p_0 \cdot x - \frac{1}{2}p_0 \cdot x_0)}\psi(x + x_0)$$

and hence

$$Q\psi(z_0) = \int_{\mathbb{R}^{2n}} e^{-\frac{i}{\hbar}p_0(x-y)}K(x,y)\psi(x+x_0)\overline{\psi(y+x_0)}dxdy. \qquad (8.5)$$

On the other hand, for every $\theta \in \mathcal{S}'(\mathbb{R}^{2n})$ we have

$$(W\psi * \theta)(z_0) = \int_{\mathbb{R}^{2n}} W\psi(z_0 - z)\theta(z)dz$$

hence, in view of the definition of the Wigner transform,

$$(W\psi * \theta)(z_0) = \left(\frac{1}{2\pi\hbar}\right)^n \int_{\mathbb{R}^{3n}} e^{-\frac{i}{\hbar}(p_0-p)\cdot y'}\psi\left(x_0 - x' + \frac{1}{2}y'\right)$$

$$\times \overline{\psi\left(x_0 - x' - \frac{1}{2}y'\right)}\theta(x',p')\,dpdx'dy'.$$

Calculating the integral in $p$ we get

$$(W\psi * \theta)(z_0) = \left(\frac{1}{2\pi\hbar}\right)^{n/2} \int_{\mathbb{R}^{3n}} F_2^{-1}\theta(x',y')e^{-\frac{i}{\hbar}p_0 \cdot y'}\psi\left(x_0 - x' + \frac{1}{2}y'\right)$$

$$\times \overline{\psi\left(x_0 - x' - \frac{1}{2}y'\right)}\theta(x',p')\,dx'dy'$$

where $F_2^{-1}$ is the inverse Fourier transform in $y'$. Making the change of variables $x' = -\frac{1}{2}(x + y)$ and $y' = x - y$ we have $dx'dy' = dxdy$ so the equality above becomes

$$(W\psi * \theta)(z_0) = \left(\frac{1}{2\pi\hbar}\right)^{n/2} \int_{\mathbb{R}^{2n}} F_2^{-1}\theta(x, x-y)e^{-\frac{i}{\hbar}p_0\cdot(x-y)}$$

$$\times \psi(x+x_0)\overline{\psi(y+x_0)}dxdy.$$

Comparison with (8.5) shows that $Q\psi = W\psi * \theta$ where $\theta$ is defined by the partial inverse Fourier transform

$$K(x, y) = \left(\frac{1}{2\pi\hbar}\right)^{n/2} F_2^{-1}\theta(x, x-y)$$

and hence

$$\theta(x, p) = (2\pi\hbar)^{n/2} \int_{\mathbb{R}^n} e^{-\frac{i}{\hbar}p\cdot y} K(x, x-y)dy. \qquad \square$$

### 8.1.1. The marginal conditions

We are mainly interested in quasi-distributions having the correct marginal properties, reproducing those of the Wigner distribution. This motivates the following terminology.

**Definition 96.** An element $Q$ of the Cohen class is said to satisfy the marginal conditions if we have, for $\psi \in L^1(\mathbb{R}^n) \cap L^2(\mathbb{R}^n)$,

$$\int_{\mathbb{R}^n} Q\psi(z)dp = |\psi(x)|^2 \quad \text{and} \quad \int_{\mathbb{R}^n} Q\psi(z)dx = |\widehat{\psi}(p)|^2. \qquad (8.6)$$

Not every element of the Cohen class satisfies the marginal conditions: see Example 98 below. When the marginal conditions are satisfied, $Q$ plays, as does the Wigner transform, the role of a quasi-distribution:

$$\int_{\mathbb{R}^{2n}} Q\psi(z)dz = 1$$

when $\|\psi\|_{L^2} = 1$.

Let us prove a necessary and sufficient condition for $Q$ to satisfy the marginal properties.

**Proposition 97.** *Let $\psi \in \mathcal{S}(\mathbb{R}^n)$ and set $Q\psi = W\psi * \theta$. Assume that the mappings $x \longmapsto \theta(x, p)$ and $p \longmapsto \theta(x, p)$ are integrable. Then we have the following properties.*

(i) *We have*

$$\int_{\mathbb{R}^n} Q\psi(z)dp = (|\psi|^2 * \alpha)(x), \quad \int_{\mathbb{R}^n} Q\psi(z)dx = (|F\psi|^2 * \beta)(p), \qquad (8.7)$$

*where the functions $\alpha$ and $\beta$ are defined by*

$$\alpha(x) = \int_{\mathbb{R}^n} \theta(x,p)dp, \quad \beta(p) = \int_{\mathbb{R}^n} \theta(x,p)dx.$$

(ii) *The marginal properties are satisfied if and only if the Fourier transform $F\theta$ exists and is such that:*

$$F\theta(x,0) = F\theta(0,p) = \left(\frac{1}{2\pi\hbar}\right)^n. \tag{8.8}$$

**Proof.** (i) In view of the first marginal property (2.26) in Chapter 2 satisfied by the Wigner distribution, we have

$$\int_{\mathbb{R}^n} W\psi(x-x',p-p')dp = |\psi(x-x')|^2$$

and hence, using Fubini's theorem,

$$\int_{\mathbb{R}^n} Q\psi(z)dp = \int_{\mathbb{R}^n} \left( \int_{\mathbb{R}^{2n}} W\psi(z-z')\theta(z')dz' \right) dp$$

$$= \int_{\mathbb{R}^{2n}} \left( \int_{\mathbb{R}^n} W\psi(z-z')dp \right) \theta(z')dz'$$

$$= \int_{\mathbb{R}^n} |\psi(x-x')|^2 \left( \int_{\mathbb{R}^n} \theta(z')dp' \right) dx',$$

which yields the first formula (8.7). The second formula is proven in a similar way using the second marginal property (2.27) of the Wigner transform.

(ii) It suffices to show that the conditions (8.8) imply that $\alpha(x) = \delta(x)$ and $\beta(p) = \delta(p)$. Let $\widehat{\theta}$ be the usual Fourier transform of the kernel $\theta$; we have

$$F\theta(x,0) = \left(\frac{1}{2\pi\hbar}\right)^n \int_{\mathbb{R}^{2n}} e^{-\frac{i}{\hbar}xx'}\theta(x',p')dp'dx'$$

$$= \left(\frac{1}{2\pi\hbar}\right)^n \int_{\mathbb{R}^n} e^{-\frac{i}{\hbar}xx'} \left( \int_{\mathbb{R}^n} \theta(x',p')dp' \right) dx'$$

$$= \left(\frac{1}{2\pi\hbar}\right)^n \int_{\mathbb{R}^n} e^{-\frac{i}{\hbar}xx'}\alpha(x')dx'.$$

hence the condition $F\theta(x,0) = (2\pi\hbar)^{-n}$ is equivalent to $\alpha(x) = \delta(x)$. Similarly, we have $F\theta(0,p) = (2\pi\hbar)^{-n}$. □

**Example 98.** The Husimi distribution considered in Example 94 does not satisfy the marginal conditions, because the functions $F\theta(x,0)$ and $F\theta(0,p)$ are not constant (they are themselves Gaussians).

### 8.1.2. Generalization of Moyal's identity

An important property of the cross-Wigner transform is the Moyal identity studied in Chapter 6. A similar identity holds for $Q$ provided that the Fourier transform of the Cohen kernel satisfies a simple condition.

**Proposition 99.** *Let $Q$ be an element of the Cohen class. We have the Moyal identity*

$$(Q(\psi,\phi)|Q(\psi',\phi'))_{L^2(\mathbb{R}^{2n})} = \left(\frac{1}{2\pi\hbar}\right)^n (\psi|\psi')_{L^2}\overline{(\phi|\phi')_{L^2}} \qquad (8.9)$$

*or, equivalently,*

$$(Q(\psi,\phi)|Q(\psi',\phi'))_{L^2(\mathbb{R}^{2n})} = (W(\psi,\phi)|W(\psi',\phi'))_{L^2(\mathbb{R}^{2n})} \qquad (8.10)$$

*if and only the Fourier transform $F\theta$ satisfies*

$$|F\theta(z)| = (2\pi\hbar)^{-n}. \qquad (8.11)$$

**Proof.** Formulas (8.9) and (8.10) are equivalent in view of Moyal's identity

$$(W(\psi,\phi)|W(\psi',\phi'))_{L^2(\mathbb{R}^{2n})} = \left(\frac{1}{2\pi\hbar}\right)^n (\psi|\psi')_{L^2}\overline{(\phi|\phi')_{L^2}}$$

(formula (6.1) in Chapter 6). Let us prove that the identity (8.9) holds if and only if $|F\theta(z)| = (2\pi\hbar)^{2n}$. Writing $Q(\psi,\phi) = W(\psi,\phi) * \theta$ and using Plancherel's formula together with the equality

$$FQ(\psi,\phi) = F(W\psi * \theta) = (2\pi\hbar)^n FW(\psi,\phi)F\theta,$$

we have, for all pairs of functions $(\psi,\phi)$ and $(\psi',\phi')$ in $\mathcal{S}(\mathbb{R}^n)$,

$$(Q(\psi,\phi)|Q(\psi',\phi'))_{L^2(\mathbb{R}^{2n})} = (2\pi\hbar)^{2n}(FW(\psi,\phi)F\theta(z)|$$
$$\times FW(\psi',\phi')F\theta(z))_{L^2(\mathbb{R}^{2n})}$$

and hence

$$(Q(\psi,\phi)|Q(\psi',\phi'))_{L^2(\mathbb{R}^{2n})} = (2\pi\hbar)^{2n}(FW(\psi,\phi)$$

$$\times |FW(\psi',\phi')|F\theta(z)|^2)_{L^2(\mathbb{R}^{2n})}.$$

The equality (8.9) is thus equivalent to $|F\theta(z)|^2 = (2\pi\hbar)^{2n}$ which proves our claim. $\qquad\square$

**Remark 100.** Since condition (8.11) implies the conditions (8.8), any element of the Cohen class which satisfies Moyal's identity also satisfies the marginal conditions (8.7).

### 8.1.3. The operator calculus associated with $Q$

Recall from Chapter 4 the Weyl operator $\widehat{A} = \mathrm{Op_W}(a)$ with symbol $a \in \mathcal{S}(\mathbb{R}^{2n})$ can be defined by the relation

$$(\widehat{A}\psi|\phi)_{L^2} = \int_{\mathbb{R}^{2n}} a(z)W(\psi,\phi)(z)dz$$

for all $\psi, \phi \in \mathcal{S}(\mathbb{R}^n)$; more generally if $a \in \mathcal{S}'(\mathbb{R}^{2n})$

$$\langle \widehat{A}\psi, \overline{\phi} \rangle = \langle a, W(\psi,\phi) \rangle.$$

One can, similarly, associate to every element $Q$ of the Cohen class an operator calculus. We are following here de Gosson [41].

**Proposition 101.** *Let $a \in \mathcal{S}'(\mathbb{R}^{2n})$ and $Q$ be an element of the Cohen such that*

$$Q : \mathcal{S}(\mathbb{R}^n) \times \mathcal{S}(\mathbb{R}^n) \longrightarrow \mathcal{S}(\mathbb{R}^{2n}).$$

*(i) There exists a unique continuous operator $\widehat{A}_Q : \mathcal{S}(\mathbb{R}^n) \longrightarrow \mathcal{S}'(\mathbb{R}^n)$ such that*

$$\langle \widehat{A}_Q\psi, \phi^* \rangle = \langle a, Q(\psi,\phi) \rangle \tag{8.12}$$

*for all $\psi, \phi \in \mathcal{S}(\mathbb{R}^n)$.*

*(ii) Let $\theta$ be the Cohen kernel of $Q$ and set $\theta^\vee(z) = \theta(-z)$. The operator $\widehat{A}_Q$ is explicitly given by*

$$\widehat{A}_Q = \left(\frac{1}{2\pi\hbar}\right)^n \int_{\mathbb{R}^{2n}} a_\sigma(z)\widehat{T}_Q(z)dz, \tag{8.13}$$

*where*

$$\widehat{T}_Q(z) = \theta_\sigma^\vee(z)\widehat{T}(z) \tag{8.14}$$

*$\theta_\sigma^\vee$ being the symplectic Fourier transform of $\theta^\vee$.*

(iii) *Equivalently,*

$$\widehat{A}_Q = \left(\frac{1}{\pi\hbar}\right)^n \int_{\mathbb{R}^{2n}} (a * \theta^\vee)(z)\widehat{R}(z)dz. \tag{8.15}$$

**Proof.** Observe that if $\phi, \psi \in \mathcal{S}(\mathbb{R}^n)$ then $Q(\psi, \phi) \in \mathcal{S}(\mathbb{R}^{2n})$, so that the right-hand side in (8.12) is well defined when $a \in \mathcal{S}'(\mathbb{R}^{2n})$. We assume that $a \in \mathcal{S}(\mathbb{R}^{2n})$, the general case follows by duality. Since

$$\langle \widehat{A}_Q\psi, \phi^* \rangle = \int_{\mathbb{R}^{2n}} a(z)(W(\psi, \phi) * \theta(z))dz,$$

we must show that the operator

$$\widehat{A} = \left(\frac{1}{2\pi\hbar}\right)^n \int_{\mathbb{R}^{2n}} a_\sigma(z)\theta_\sigma^\vee(z)\widehat{T}(z)dz \tag{8.16}$$

is identical to $\widehat{A}_Q$, that is

$$\langle \widehat{A}_Q\psi, \phi^* \rangle = \int_{\mathbb{R}^{2n}} a(z)(W(\psi, \phi) * \theta)(z)dz$$

for all $\phi, \psi \in \mathcal{S}(\mathbb{R}^n)$. In view of (8.16) we have

$$\langle \widehat{A}_Q\psi, \phi^* \rangle = \left(\frac{1}{2\pi\hbar}\right)^n \int_{\mathbb{R}^{2n}} a_\sigma(z_0) \left( \int_{\mathbb{R}^n} \theta_\sigma(z_0)\widehat{T}(z_0)\psi(x)\phi^*(x)dx \right) dz_0$$

$$= \left(\frac{1}{2\pi\hbar}\right)^n \int_{\mathbb{R}^{2n}} a_\sigma(z_0)\theta_\sigma(z_0)\langle \phi|\widehat{T}(z_0)\psi\rangle dz_0.$$

In view of the definition (3.1) of the cross-ambiguity function,

$$\langle \phi|\widehat{T}(z_0)\psi\rangle = \langle \widehat{T}(-z_0)\phi|\psi\rangle = (2\pi\hbar)^n A(\psi, \phi)(-z_0)$$

and hence

$$\langle \widehat{A}_Q\psi, \phi^* \rangle = \int_{\mathbb{R}^{2n}} a_\sigma(z_0)\theta_\sigma^\vee(z_0)A(\psi, \phi)(-z_0)dz_0$$

$$= \int_{\mathbb{R}^{2n}} a_\sigma(z_0)(\theta_\sigma A(\psi, \phi))(-z_0)dz_0,$$

where $\theta_\sigma^\vee(z) = \theta_\sigma(-z)$. Using the Plancherel formula for the symplectic Fourier transform, and using the relation $F_\sigma a_\sigma = a$, this equality becomes

$$\langle \widehat{A}_Q \psi, \phi^* \rangle = \int_{\mathbb{R}^{2n}} a(z_0) F_\sigma(\theta_\sigma A(\psi, \phi))(z_0) dz_0$$

$$= (2\pi\hbar)^{-n} \int_{\mathbb{R}^{2n}} a(z_0)(F_\sigma \theta_\sigma * F_\sigma A(\psi, \phi))(z_0) dz_0;$$

and using the relation $F_\sigma \theta_\sigma = \theta$ this yields

$$\langle \widehat{A}_Q \psi, \phi^* \rangle = (2\pi\hbar)^{-n} \int_{\mathbb{R}^{2n}} a(z_0)(\theta * W(\psi, \phi))(z_0) dz_0$$

$$= \int_{\mathbb{R}^{2n}} a(z_0) Q(\psi, \phi)(z_0) dz_0$$

as we set out to prove. $\qquad\square$

## 8.2. Two Examples

### 8.2.1. The generalized Husimi distribution

We consider here elements $Q\psi = W\psi * \theta$ of the Cohen class where the kernel $\theta$ is itself a Wigner transform.

**Definition 102.** An element of the Cohen class of the type $Q_\phi \psi = W\psi * W\phi$ where $\phi \in \mathcal{S}(\mathbb{R}^n)$ is called a generalized Husimi distribution.

The basic example is provided by the usual Husimi distribution.

**Example 103.** If $\phi$ is the standard Gaussian $\phi_0(x) = (\pi\hbar)^{-n/4} e^{-|x|^2/2\hbar}$, then $Q_0 = W\psi * W\phi_0$ is called the Husimi distribution; it is explicitly given by

$$Q_0 \psi(z) = (\pi\hbar)^{-n}(W\psi * e^{-|\cdot|^2/\hbar})(z)$$

in view of formula (9.16) in Chapter 9.

The generalized Husimi distribution does not satisfy the marginal properties (cf. Example 94); it is however positive.

**Proposition 104.** *Let* $\psi, \phi \in \mathcal{S}(\mathbb{R}^n)$. *We have* $Q_\phi \psi \geq 0$; *more precisely*

$$Q_\phi \psi(z) = |A(\psi, \phi^\vee)(z)|^2 \geq 0, \qquad (8.17)$$

*where* $\phi^\vee(x) = \phi(-x)$ *and* $A(\psi, \phi^\vee)$ *is the cross-ambiguity transform of* $(\psi, \phi^\vee)$. *Equivalently:*

$$Q_\phi \psi(z) = 2^n \left| W(\psi, \phi) \left( \frac{1}{2} z \right) \right|^2 \geq 0. \qquad (8.18)$$

**Proof.** In view of formula (3.17) in Chapter 3, we have

$$A(\psi, \phi)(z) = 2^{-n} W(\psi, \phi^\vee) \left( \frac{1}{2} z \right)$$

hence the equivalence of (8.17) and (8.18). Let us prove (8.17). Observing that $W(\phi^\vee) = (W\phi)^\vee$ we have

$$Q_\phi \psi(z) = \int_{\mathbb{R}^{2n}} W\psi(z - z') W\phi(z') dz'$$

$$= \int_{\mathbb{R}^{2n}} W\psi(z + z') W\phi(-z') dz'$$

$$= \int_{\mathbb{R}^{2n}} W\psi(z + z') W\phi^\vee(z') dz',$$

that is, since $W\psi(z + z') = W(\widehat{T}(-z)\psi)$ and $W\phi^\vee$ is real,

$$Q_\phi \psi(z) = \int_{\mathbb{R}^{2n}} W\widehat{T}(-z)\psi(z') \overline{W(\phi^\vee)}(z') dz'$$

$$= (W(\widehat{T}(-z)\psi), W(\phi^\vee))_{L^2(\mathbb{R}^n)}.$$

In view of the Moyal identity (6.1) in Chapter 6, this equality can be rewritten as

$$Q_\phi \psi(z) = \left( \frac{1}{2\pi\hbar} \right)^n |(\widehat{T}(-z)\psi, \phi^\vee)_{L^2}|^2$$

$$= \left( \frac{1}{2\pi\hbar} \right)^n |(\psi, \widehat{T}(z)\phi^\vee)_{L^2}|^2$$

hence $Q_\phi \psi \geq 0$ and formula (8.17) follows by definition of the cross-ambiguity function. $\square$

**Remark 105.** Let $\Phi_R(z) = e^{-|z|^2/R^2}$. De Bruijn has proved in [14] that the convolution $W\psi * \Phi_R$ satisfies

$$W\psi * \Phi_R \geq 0 \text{ if } R^2 = \hbar, \quad W\psi * \Phi_R > 0 \text{ if } R^2 > \hbar.$$

We will return to this fact when we discuss the notion of "quantum blob" in Chapter 15.

Recall (formula (8.13)) that to every element

$$Q : \mathcal{S}(\mathbb{R}^n) \times \mathcal{S}(\mathbb{R}^n) \longrightarrow \mathcal{S}(\mathbb{R}^{2n})$$

of the Cohen class with kernel $\theta$, one associates an operator

$$\widehat{A}_Q = \left(\frac{1}{2\pi\hbar}\right)^n \int_{\mathbb{R}^{2n}} a_\sigma(z)\widehat{T}_Q(z)dz, \tag{8.19}$$

where $\widehat{T}_Q(z) = \theta_\sigma^\vee(z)\widehat{T}(z)$. In the present case $\theta = W\phi$ for some $\phi \in \mathcal{S}(\mathbb{R}^n)$ hence $\theta_\sigma^\vee(z) = A\phi(-z)$ so that the corresponding operator (8.19) is explicitly given by

$$\widehat{A}_\phi = \left(\frac{1}{2\pi\hbar}\right)^n \int_{\mathbb{R}^{2n}} a_\sigma(z)A\phi(-z)\widehat{T}(z)dz.$$

In view of the convolution formula

$$F_\sigma(a * b) = (2\pi\hbar)^n (F_\sigma a)(F_\sigma a)$$

(see Appendix B), we can also write

$$\widehat{A}_\phi = \left(\frac{1}{2\pi\hbar}\right)^n \int_{\mathbb{R}^{2n}} a_\sigma(z)F_\sigma\theta^\vee(z)\widehat{T}(z)dz$$

$$= \left(\frac{1}{2\pi\hbar}\right)^{2n} \int_{\mathbb{R}^{2n}} F_\sigma(a * \theta^\vee)(z)\widehat{T}(z)dz$$

and hence

$$\widehat{A}_\phi = \left(\frac{1}{2\pi\hbar}\right)^n \mathrm{Op_W}(a * \theta^\vee).$$

**Example 106.** Choose $\phi_0(x) = (\pi\hbar)^{-n/4}e^{-|x|^2/2\hbar}$ (the standard Gaussian). Then $W\phi(z) = (\pi\hbar)^{-n}e^{-\frac{1}{\hbar}|z|^2}$ and hence

$$a * \theta^\vee(z) = (\pi\hbar)^{-n} \int_{\mathbb{R}^{2n}} e^{-\frac{1}{\hbar}|z-z'|^2}a(z')dz'.$$

The operator $\widehat{A}_{\phi_0}$ is the anti-Wick operator with symbol $a$ (see [7, 39, 76] for studies of these operators).

### 8.2.2. The Born–Jordan transform

Let us introduce the *sine cardinal* function familiar from signal analysis: by definition

$$\mathrm{sinc}(t) = \frac{\sin t}{t} \quad \text{if } t \neq 0, \quad \mathrm{sinc}\, 0 = 1.$$

**Definition 107.** The Born–Jordan distribution is the element $W_{\mathrm{BJ}}$ of the Cohen class corresponding to the kernel

$$\theta_{\mathrm{BJ}} = \left(\frac{1}{2\pi\hbar}\right)^n F\chi_{\mathrm{BJ}},$$

where the function $\chi_{\mathrm{BJ}}$ is defined by

$$\chi_{\mathrm{BJ}}(x,p) = \mathrm{sinc}\left(\frac{p \cdot x}{2\hbar}\right).$$

Noting that since $F\chi_{\mathrm{BJ}} = F_\sigma \chi_{\mathrm{BJ}}$ because the function since being even, the symplectic Fourier transform

$$A_{\mathrm{BJ}}(\psi,\phi) = F_\sigma W_{\mathrm{BJ}}(\psi,\phi)$$

of the Born–Jordan distribution is given by

$$A_{\mathrm{BJ}}(\psi,\phi)(x,p) = A(\psi,\phi)(x,p)\mathrm{sinc}\left(\frac{p \cdot x}{2\hbar}\right),$$

where $A(\psi,\phi)$ is the usual cross-ambiguity function. Since multiplication by $\mathrm{sinc}(px/2\hbar)$ is easier to handle than convolution by the Fourier transform of that function, it is often preferable to work with $A_{\mathrm{BJ}}(\psi,\phi)$ than with $W_{\mathrm{BJ}}(\psi,\phi)$.

The Born–Jordan distribution satisfies the marginal conditions:

$$\int_{\mathbb{R}^n} W_{\mathrm{BJ}}(\psi,\phi)(z)dp = |\psi(x)|^2,$$

$$\int_{\mathbb{R}^n} W_{\mathrm{BJ}}(\psi,\phi)(z)dx = |F\psi(p)|^2.$$

In fact, by definition of $\theta_{\mathrm{BJ}}$ we have

$$F\theta_{\mathrm{BJ}}(x,p) = \left(\frac{1}{2\pi\hbar}\right)^n \mathrm{sinc}\left(\frac{p \cdot x}{2\hbar}\right)$$

and hence $F\theta_{\mathrm{BJ}}(x,0) = F\theta_{\mathrm{BJ}}(0,p) = (2\pi\hbar)^{-n}$ which is condition (8.8).

The Born–Jordan distribution $W_{\mathrm{BJ}}$ does not satisfy the generalized Moyal identity

$$(Q(\psi,\phi)|Q(\psi',\phi'))_{L^2(\mathbb{R}^{2n})} = \left(\frac{1}{2\pi\hbar}\right)^n (\psi|\psi')_{L^2}(\phi|\phi')_{L^2}$$

since condition (8.11) is not satisfied.

The operators (8.19) corresponding to the kernel $\theta_{\mathrm{BJ}}$ are called the *Born–Jordan operators*; one writes $\widehat{A}_{\mathrm{BJ}} = \mathrm{Op}_{\mathrm{BJ}}(a)$ and we have by definition

$$\widehat{A}_{\mathrm{BJ}} = \left(\frac{1}{2\pi\hbar}\right)^n \int_{\mathbb{R}^{2n}} a_\sigma(z)\mathrm{sinc}\left(\frac{p\cdot x}{2\hbar}\right)\widehat{T}(z)dz. \tag{8.20}$$

The terminology comes from the fact that in the case of monomial symbols $x_j^r p_j^s$ we have

$$\mathrm{Op}_{\mathrm{BJ}}(x_j^r p_j^s) = \frac{1}{r+1}\sum_{k=0}^r \widehat{x}_j^{\,r-k}\widehat{p}_j^{\,s}\widehat{x}_j^{\,jk} \tag{8.21}$$

which is the monomial quantization rule proposed by the physicists Born and Jordan [12] in 1925, and which predates the Weyl rule

$$\mathrm{Op}_{\mathrm{W}}(x_j^r p_j^s) = \frac{1}{2^s}\sum_{k=0}^s \binom{s}{k}\widehat{x}_j^{\,s-k}\widehat{p}_j^{\,r}\widehat{x}_j^{\,k} \tag{8.22}$$

(formula (4.7) in Chapter 4); both rules coincide only when $r+s \le 2$ (hence, in particular, they lead to the same quantization of quadratic polynomials).

One proves [8–10, 41] that the Born–Jordan quantization can be obtained by an averaging procedure on certain elements of the Cohen class: for $\tau \in \mathbb{R}$ define the $\tau$-cross-Wigner transform by

$$W_\tau(\psi,\phi)(z) = \left(\frac{1}{2\pi\hbar}\right)^n \int_{\mathbb{R}^n} e^{-\frac{i}{\hbar}py}\psi(x+\tau y)\overline{\phi(x-(1-\tau)y)}dy; \tag{8.23}$$

it reduces to the ordinary cross-Wigner transform when $\tau = \frac{1}{2}$. One proves that

$$W_\tau(\psi,\phi) = W(\psi,\phi) * \theta_{(\tau)}, \tag{8.24}$$

where the Cohen kernel $\theta_{(\tau)}$ is the function

$$\theta_{(\tau)}(z) = \frac{2^n}{|2\tau-1|^n}e^{\frac{2i}{\hbar(2\tau-1)}px} \tag{8.25}$$

for $\tau \neq \frac{1}{2}$ and $\theta_{(1/2)} = \delta$. Integrating the equality (8.24) with respect to $\tau \in [0, 1]$, one gets

$$\int_0^1 W_\tau(\psi, \phi)(z) d\tau = W_{\mathrm{BJ}}(\psi, \phi)(z).$$

There are strong indications that the Born–Jordan quantization procedure, given in the general case by formula (8.20), might be the physically correct quantization scheme, as opposed to the Weyl quantization. See [15, 41, 42] for detailed studies of the Born–Jordan operators.

## Main Formulas in Chapter 8

| | |
|---|---|
| Cohen class | $Q(\psi, \phi) = W(\psi, \phi) * \theta$ |
| Cohen class (bis) | $F_\sigma Q(\psi, \phi) = (2\pi\hbar)^n A(\psi, \phi) F_\sigma \theta$ |
| Husimi distribution | $\theta(z) = (\pi\hbar)^{-n} e^{-\frac{1}{\hbar}|z|^2}$ |
| Marginal properties hold if | $F\theta(x, 0) = F\theta(0, p) = \left(\frac{1}{2\pi\hbar}\right)^n$ |
| Moyal identity holds if | $|F\theta(z)| = (2\pi\hbar)^{-n}$ |
| Operator $\widehat{A}_Q$ | $\langle \widehat{A}_Q \psi, \phi^* \rangle = \langle a, Q(\psi, \phi) \rangle$ |
| Equivalently | $\widehat{A}_Q = \left(\frac{1}{2\pi\hbar}\right)^n \int_{\mathbb{R}^{2n}} a_\sigma(z) \theta_\sigma^\vee(z) \widehat{T}(z) dz$ |
| Born–Jordan case | $\theta_{\mathrm{BJ}} = \left(\frac{1}{2\pi\hbar}\right)^n F\chi_{\mathrm{BJ}}$ , $\chi_{\mathrm{BJ}}(z) = \mathrm{sinc}\left(\frac{px}{2\hbar}\right)$ |
| Born–Jordan operator | $\widehat{A}_{\mathrm{BJ}} = \left(\frac{1}{2\pi\hbar}\right)^n \int_{\mathbb{R}^{2n}} a_\sigma(z) \mathrm{sinc}\left(\frac{px}{2\hbar}\right) \widehat{T}(z) dz$ |

# Chapter 9

# Gaussians and Hermite Functions

**Summary 108.** We compute the cross-Wigner transform and the cross-ambiguity function of pairs of Gaussian functions. The Wigner transform of a function is positive if and only if that function is a (generalized) Gaussian: this is Hudson's theorem, later generalized by Janssen. We thereafter study the cross-Wigner transform of Hermite functions.

We begin by studying the cross-Wigner transform of generalized Gaussians; these are the "squeezed coherent states" appearing in quantum mechanics and optics. We thereafter discuss in detail the case of Hermite functions.

## 9.1. Wigner Transform of Generalized Gaussians

### 9.1.1. Generalized Gaussian functions

In what follows $X$ and $Y$ are real symmetric $n \times n$ matrices; we assume in addition that $X$ is positive definite (and hence invertible): $X > 0$. We set

$$M = X + iY;$$

it is a complex symmetric $n \times n$ matrix. To $M$ we associate the (complex) generalized Gaussian function

$$\psi_M(x) = \left( \frac{1}{\pi \hbar} \right)^{n/4} (\det X)^{1/4} e^{-\frac{1}{2\hbar} M x^2}, \tag{9.1}$$

where $Mx^2 = Mx \cdot x$ (complex scalar product). The coefficient in front of the exponential is chosen so that $\psi_M$ is normalized to unity: $||\psi_M||_{L^2} = 1$.

**Example 109.** The standard Gaussian (also called "fiducial coherent state" in the physical literature) is defined by

$$\phi_0(x) = (\pi\hbar)^{-n/4} e^{-|x|^2/2\hbar}. \tag{9.2}$$

It corresponds to the choice $M = I_{\mathrm{d}}$.

Functions of the type (9.1) are called *squeezed coherent states* in the physical literature. They can be obtained by transforming the fiducial coherent state (9.2) by metaplectic operators.

In the case $n = 1$ the standard Gaussian (9.2) is a solution of the differential equation

$$\frac{1}{2}\left(-\hbar^2\partial_x^2 + x^2\right)\phi_0 = \frac{1}{2}\hbar\phi_0 \tag{9.3}$$

(when $\hbar = 1$ this the "Hermite equation"). A natural question to ask is whether the generalized Gaussians (9.1) also solve some second-order differential equation. This is indeed the case; to determine that equation we will use the following elementary result, the proof of which is calculatory, and left to the reader.

**Lemma 110.** *Let $\psi$ be a twice continuously differentiable function defined on $\mathbb{R}^n$. Writing $\psi(x) = R(x)e^{i\Phi(x)/\hbar}$ where $R \geq 0$ and $\Phi$ are real, that function satisfies the equation $\widehat{H}\psi = 0$ where*

$$\widehat{H} = (-i\hbar\nabla_x - \partial_x\Phi)^2 + \hbar^2\frac{\Delta R}{R} \tag{9.4}$$

*at all points $x$ where $R(x) > 0$.*

Choosing in particular $\psi = \psi_M$ we obtain the following proposition.

**Proposition 111.** *The generalized Gaussian function*

$$\psi_M(x) = \left(\frac{1}{\pi\hbar}\right)^{n/4} (\det X)^{1/4} e^{-\frac{1}{2\hbar}Mx^2}$$

*($M = X + iY, X > 0$) satisfies the eigenvalue problem*

$$\widehat{H}_M\psi_M = (\hbar\,\mathrm{Tr}X)\psi_M, \tag{9.5}$$

*where $\widehat{H}_M$ is the partial differential operator*

$$\widehat{H}_M = (-i\hbar\partial_x + Yx)^2 + X^2x \cdot x.$$

**Proof.** Applying Lemma 110 with $\Phi(x) = -\frac{1}{2}Yx \cdot x$ and $R(x) = e^{-\frac{1}{2\hbar}Xx \cdot x}$, we get

$$\partial_x \Phi(x) = -Yx, \quad \frac{\Delta R(x)}{R(x)} = -\frac{1}{\hbar}\operatorname{Tr}X + \frac{1}{\hbar^2}X^2 x \cdot x \qquad (9.6)$$

hence the operator $\widehat{H}$ defined by (9.4) is given by

$$\widehat{H} = (-i\hbar\partial_x + Yx)^2 - \hbar\operatorname{Tr}X + X^2 x \cdot x;$$

formula (9.5) follows since the Weyl symbol of $\widehat{H}$ is the function $H_M - \hbar\operatorname{Tr}X$. □

In physical terminology one would say that squeezed coherent states are eigenfunctions of generalized harmonic oscillators.

### 9.1.2. Explicit results

We now calculate the cross-Wigner transform of a pair of generalized Gaussians. We will use the following generalized Fresnel formula (see, e.g. [28]): let

$$\phi_M(x) = e^{-\frac{1}{2\hbar}Mx^2},$$

that is

$$\phi_M = (\pi\hbar)^{n/4}(\det X)^{-1/4}\psi_M.$$

We have

$$F\phi_M(x) = (\det M)^{-1/2}\phi_{M^{-1}}(x), \qquad (9.7)$$

where $(\det M)^{1/2} = \lambda_1^{1/2}\lambda_2^{1/2}\cdots\lambda_n^{1/2}$ the $\lambda_j^{1/2}$ being the square roots with positive real parts of the eigenvalues of $M$.

**Proposition 112.** *Let $\psi_M$ and $\psi_{M'}$ be of the type (9.1). We have*

$$W(\psi_M, \psi_{M'})(z) = \left(\frac{1}{\pi\hbar}\right)^n C_{M,M'} e^{-\frac{1}{\hbar}Fz^2} \qquad (9.8)$$

*where $C_{M,M'}$ is the constant*

$$C_{M,M'} = (\det XX')^{1/4}\det\left[\frac{1}{2}(M + \overline{M'})\right]^{-1/2} \qquad (9.9)$$

*and $F$ is the symmetric complex matrix given by*

$$F = \begin{pmatrix} 2\overline{M'}(M + \overline{M'})^{-1}M & -i(M - \overline{M'})(M + \overline{M'})^{-1} \\ -i(M + \overline{M'})^{-1}(M - \overline{M'}) & 2(M + \overline{M'})^{-1} \end{pmatrix}. \quad (9.10)$$

**Proof.** By definition of the cross-Wigner transform, we have

$$W(\psi_M, \psi_{M'})(z) = C(X, X') \int_{\mathbb{R}^n} e^{-\frac{i}{\hbar}py} e^{-\frac{1}{2\hbar}\Phi(x,y)} dy,$$

where the functions $C(X, X')$ and $\Phi(x, y)$ are given by

$$C(X, X') = 2^{-n} \left(\frac{1}{\pi\hbar}\right)^{2n} (\det XX')^{1/4},$$

$$\Phi(x, y) = M\left(x + \frac{1}{2}y\right)^2 + \overline{M'}\left(x - \frac{1}{2}y\right)^2.$$

Let us evaluate the integral

$$I(x, p) = \int_{\mathbb{R}^n} e^{-\frac{i}{\hbar}p \cdot y} e^{-\frac{1}{2\hbar}\Phi(x,y)} dy.$$

We have

$$\Phi(x, y) = (M + \overline{M'})x^2 + \frac{1}{4}(M + \overline{M'})y^2 + (M - \overline{M'})x \cdot y$$

and hence

$$I(x, p) = e^{-\frac{1}{2\hbar}(M+\overline{M'})x^2} \int_{\mathbb{R}^n} e^{-\frac{i}{\hbar}[p - \frac{i}{2}(M - \overline{M'})x] \cdot y} e^{-\frac{1}{8\hbar}(M+\overline{M'})y^2} dy.$$

Using the "Fresnel formula" (9.7) we get

$$I(x, p) = (2\pi\hbar)^{n/2} \det\left[\frac{1}{4}(M + \overline{M'})\right]^{-1/2} \times \exp\left(-\frac{1}{2\hbar}\left[(M + \overline{M'})x^2\right.\right.$$

$$\left.\left. + 4(M + \overline{M'})^{-1}\left(p - \frac{1}{2}(M - \overline{M'})x\right)^2\right]\right).$$

A straightforward calculation shows that

$$\frac{1}{2}(M + \overline{M'})x^2 + 4(M + \overline{M'})^{-1}\left(p - \frac{1}{2}(M - \overline{M'})x\right)^2 = Fz \cdot z,$$

where

$$F = \begin{pmatrix} K & -i(M - \overline{M'})(M + \overline{M'})^{-1} \\ -i(M + \overline{M'})^{-1}(M - \overline{M'}) & 2(M + \overline{M'})^{-1} \end{pmatrix} \quad (9.11)$$

with

$$K = \frac{1}{2}\left[M + \overline{M'} - (M - \overline{M'})(M + \overline{M'})^{-1}(M - \overline{M'})\right].$$

Using the algebraic identity

$$M + \overline{M'} - (M - \overline{M'})(M + \overline{M'})^{-1}(M - \overline{M'}) = 4\overline{M'}(M + \overline{M'})^{-1}M$$

$$(9.12)$$

the matrix (9.11) is given by (9.10). Summarizing, we get after some simplifications

$$W(\psi_M, \psi_{M'})(z) = \left(\frac{1}{\pi\hbar}\right)^n (\det XX')^{1/4} \det\left[\frac{1}{2}(M + \overline{M'})\right]^{-1/2} e^{-\frac{1}{\hbar}Fz^2}$$

which proves (9.8). $\square$

The following particular case of the result above yields the well-known Wigner transform of a Gaussian.

**Corollary 113.** *The Wigner transform of the generalized Gaussian (9.1) is the phase space Gaussian*

$$W\psi_M(z) = \left(\frac{1}{\pi\hbar}\right)^n e^{-\frac{1}{\hbar}Gz^2}, \quad (9.13)$$

*where G is the symmetric matrix*

$$G = \begin{pmatrix} X + YX^{-1}Y & YX^{-1} \\ X^{-1}Y & X^{-1} \end{pmatrix}. \quad (9.14)$$

*We have $G \in \mathrm{Sp}(n)$. In fact $G = S^T S$ where*

$$S = \begin{pmatrix} X^{1/2} & 0 \\ X^{-1/2}Y & X^{-1/2} \end{pmatrix}. \quad (9.15)$$

**Proof.** Setting $M = M'$ we have $M + \overline{M'} = 2X$ and $M + \overline{M'} = 2iY$ and hence

$$2\overline{M'}(M + \overline{M'})^{-1}M = (X - iY)X^{-1}(X + iY) = X + YX^{-1};$$

similarly

$$-i(M - \overline{M'})(M + \overline{M'})^{-1} = YX^{-1} \ and \ 2(M + \overline{M'})^{-1} = X^{-1}$$

hence $F = G$.                                                                              □

**Remark 114.** Serge de Gosson (private communication) has shown that the matrix $F$ in (9.11) is complex symplectic: $F^T J F = J$.

A basic but yet useful example is as follows.

**Example 115.** Consider the standard Gaussian

$$\phi_0(x) = (\pi\hbar)^{-n/4} e^{-|x|^2/2\hbar}.$$

Application of (9.13) immediately yields:

$$W\phi_0(z) = (\pi\hbar)^{-n} e^{-\frac{1}{\hbar}|z|^2}. \tag{9.16}$$

### 9.1.3. Cross-ambiguity function of a Gaussian

The calculation of cross-ambiguity functions of Gaussians is straightforward if one uses the algebraic relation (3.17) in Proposition 33 of Chapter 3:

$$A(\psi, \phi)(z) = 2^{-n} W(\psi, \phi^\vee)(\frac{1}{2}z),$$

where $\phi^\vee(x) = \phi(-x)$.

**Proposition 116.** *Let $\psi_M$ and $\psi_{M'}$ be generalized Gaussian* (9.1). *We have*

$$A(\psi_M, \psi_{M'})(z) = \left(\frac{1}{2\pi\hbar}\right)^n C_{M,M'} e^{-\frac{1}{4\hbar}Fz^2}, \tag{9.17}$$

*where the constant $C_{M,M'}$ and the matrix $F$ are again given by* (9.9) *and* (9.10). *In particular*

$$A\psi_M(z) = \left(\frac{1}{2\pi\hbar}\right)^n e^{-\frac{1}{4\hbar}Gz^2}$$

*where the symplectic matrix $G$ is given by* (9.14).

**Proof.** The proof of this result is obvious, taking into account the fact that generalized Gaussians are even functions.                              □

**Remark 117.** These formulas can also be derived using the fact that the cross-ambiguity function is the symplectic Fourier transform of the cross-Wigner transform. Evidently, the calculations are more complicated.

**Example 118.** Let $\phi_0(x) = (\pi\hbar)^{-n/4}e^{-|x|^2/2\hbar}$ be the standard Gaussian. We have

$$A\phi_0(z) = (2\pi\hbar)^{-n}e^{-\frac{1}{4\hbar}|z|^2}.$$

### 9.1.4. Hudson's theorem

Hudson showed in [59] that the Wigner transform of a function on $\mathbb{R}$ is positive if and only if that function is a generalized Gaussian. Janssen generalized in [60] Hudson's result to the multidimensional case under more general hypotheses. See [28, 52, 77, 79, Chapter 1] for independent proofs.

In our notation and terminology this result can be restated as follows.

**Proposition 119.** *Let $\psi \in L^2(\mathbb{R}^n)$. We have $W\psi > 0$ if and only if there exist $M$ and $z_0$ such that $\psi = T(z_0)\psi_M$.*

Since all the known approaches involve at some point the use of techniques from the theory of analytic functions, which are out of the scope of this monograph, we omit the proof of this result here and refer to the texts mentioned above. We however mention the following refinement of Hudson's theorem apparently due to Toft [79].

**Proposition 120.** *Let $\psi, \phi \in L^2(\mathbb{R}^n)$. We have $W(\psi, \phi) > 0$ if and only if $\psi = T(z_0)\psi_M$ and $\phi = \lambda\psi$ for some $\lambda > 0$.*

It follows from Proposition 112 that the cross-Wigner transform of a pair of arbitrary Gaussians is not even a real function.

### 9.2. The Case of Hermite Functions

### 9.2.1. Short review of the Hermite and Laguerre functions

The following formulas are to be found widespread in the literature; we refer to the monumental treatise [48] by Gradshteyn and Ryzhik. We denote by $h_N$ the $N$-th Hermite polynomial; it is defined by "Rodrigue's formula"

$$h_N(x) = (-1)^N e^{x^2}\left(\frac{d^N}{dx^N}e^{-x^2}\right). \tag{9.18}$$

The functions $h_N$ satisfy Hermite's differential equation

$$\frac{d^2}{dx^2}h_N - 2x\frac{d}{dx}h_N + 2Nh_N = 0 \tag{9.19}$$

and the recurrence relations

$$\frac{d}{dx}h_N = 2Nh_{N-1}, \quad h_{N+1} = 2xh_N - 2Nh_{N-1}. \tag{9.20}$$

**Definition 121.** The $N$th Hermite function is given by

$$H_N(x) = \sqrt{\frac{1}{2^N N!}} \left(\frac{1}{\pi}\right)^{1/4} e^{-x^2/2} h_N(x). \tag{9.21}$$

Setting $N = 0$ we have $h_0(x) = 1$ hence

$$H_0(x) = \pi^{-1/4} e^{-x^2/2}$$

is the standard Gaussian corresponding to the choice $\hbar = 1$.

The Hermite functions satisfy the parity relation

$$H_N(-x) = (-1)^N H_N(x) \tag{9.22}$$

and they are the eigenfunctions of the "Hermite operator"

$$\widehat{H} = \frac{1}{2}\left(-\partial_x^2 + x^2\right); \tag{9.23}$$

more precisely

$$\frac{1}{2}\left(-\partial_x^2 + x^2\right)\phi_N(x) = \left(N + \frac{1}{2}\right)\phi_N(x) \tag{9.24}$$

for $N = 0, 1, 2, \ldots$.

We will use a rescaled variant of the Hermite functions which is very much used in quantum mechanics. Consider the operator

$$\widehat{H} = \frac{1}{2m}\left(-\hbar^2\partial_x^2 + m^2\omega^2 x^2\right);$$

it is the quantization of the harmonic oscillator Hamiltonian function

$$H = \frac{1}{2m}\left(p^2 + m^2\omega^2 x^2\right). \tag{9.25}$$

The operator $\widehat{H}$ has the eigenvalues $\lambda_N = (N + \frac{1}{2})\hbar\omega$ and its eigenfunctions $\Phi_N$ are given by

$$\Phi_N(x) = \sqrt{\alpha}H_N(\alpha x), \quad \alpha = \sqrt{\frac{m\omega}{\hbar}},$$

that is, in view of (9.21),

$$\Phi_N(x) = \sqrt{\frac{1}{2^N N!}}\left(\frac{m\omega}{\pi\hbar}\right)^{1/4} e^{-m\omega x^2/2\hbar} h_N\left(\sqrt{\frac{m\omega}{\hbar}}x\right). \tag{9.26}$$

Let us next recall the basics of the theory of Laguerre functions.

**Definition 122.** The $N$th generalized Laguerre polynomial $L_N^{(k)}(k \in \mathbb{R})$ is given by Rodrigue's formula

$$L_N^{(k)}(x) = \frac{e^x x^{-k}}{N!} \frac{d^N}{dx^N}(x^{N+k} e^{-x}).$$ (9.27)

The $N$th (simple) Laguerre polynomial is $L_N = L_N^{(0)}$, that is,

$$L_N(x) = e^x \frac{d^N}{dx^N}(x^N e^{-x}).$$ (9.28)

Applying Leibniz's rule for the differentiation of a product to Rodrigue's' formulas (9.27) and (9.28), we have the explicit expression

$$L_N^{(k)}(x) = \sum_{j=0}^{N} \frac{1}{j!} \binom{N+k}{N-j}(-x)^j$$

and hence

$$L_N(x) = \sum_{j=0}^{N} \binom{N}{j} \frac{(-x)^j}{j!}.$$

The Laguerre polynomials satisfy the Laguerre equation

$$\frac{d^2}{dx^2} L_N^{(k)} + (k+1-x)\frac{d}{dx}L_N^{(k)} + NL_N^{(k)} = 0.$$

We have the following orthogonality relations:

$$\int_0^\infty x^k e^{-x} L_M^{(k)}(x) L_N^{(k)}(x)dx = \frac{\Gamma(N+k+1)}{N!}\delta_{M,N}.$$

In particular:

$$\int_0^\infty e^{-x} L_M(x) L_N(x)dx = \delta_{M,N}.$$

The following integral is very useful (see [48, 7.377]): if $M \leq N$ then

$$\int_{-\infty}^{\infty} e^{-x^2} h_M(x+u) h_N(x+v)dx = 2^n \sqrt{\pi} M! v^{N-M} L_M^{(N-M)}(-2uv);$$ (9.29)

choosing $M = N$ we have

$$\int_{-\infty}^{\infty} e^{-x^2} h_N(x+u) h_N(x+v)dx = 2^n \sqrt{\pi} N! L_N(-2uv).$$ (9.30)

### 9.2.2. The Wigner transform of Hermite functions

We consider quantum mechanical Hermite functions. The result is well known, and one can find variants of the proof in many texts, for instance Dahl [16], Groenewold [54], Hillery *et al.* [58].

**Proposition 123.** *The Wigner transform of the $N$th Hermite function $\Phi_N$ is given by*

$$W\Phi_N(z) = \frac{(-1)^N}{\pi\hbar} e^{-2H(z)/\hbar\omega} L_N\left(\frac{4}{\hbar\omega}H(z)\right), \qquad (9.31)$$

*where $z = (x,p)$ and $L_N$ is the $N$-th simple Laguerre polynomial, $H$ being defined by (9.25).*

**Proof.** We have, by definition of the Wigner transform,

$$W\Phi_N(z) = \frac{1}{2\pi\hbar} \int_{-\infty}^{\infty} e^{-\frac{i}{\hbar}py}\Phi_N\left(x+\frac{1}{2}y\right)\Phi_N\left(x-\frac{1}{2}y\right)dy$$

that is, in view of (9.26),

$$W\Phi_N(z) = \frac{1}{2}C_N \int_{-\infty}^{\infty} e^{-\frac{i}{\hbar}py}e^{-\alpha^2\left(x^2+\frac{1}{4}y^2\right)}h_N$$

$$\times\left(\alpha\left(x+\frac{1}{2}y\right)\right)h_N\left(\alpha\left(x-\frac{1}{2}y\right)\right)dy$$

$$= C_N e^{-\alpha^2 x^2}\int_{-\infty}^{\infty} e^{-\frac{i}{2\hbar}py}e^{-\alpha^2 y^2}h_N(\alpha(x+y))h_N(\alpha(x-y))dy,$$

where $C_N$ and $\alpha$ are constants:

$$C_N = \frac{1}{\pi\hbar}\frac{1}{2^N N!}\left(\frac{m\omega}{\pi\hbar}\right)^{1/2}, \quad \alpha = \sqrt{\frac{m\omega}{\hbar}}.$$

Completing squares we have

$$\alpha^2 y^2 + \frac{i}{2\hbar}py = \alpha^2\left(y-\frac{ip}{\alpha^2\hbar}\right)^2 + \frac{p^2}{\alpha^2\hbar^2}$$

and hence

$$W\Phi_N(z) = C_N e^{-\alpha^2 x^2}e^{-\frac{p^2}{\alpha^2\hbar^2}}\int_{-\infty}^{\infty} e^{-\alpha^2\left(y-\frac{ip}{\alpha^2\hbar}\right)^2}h_N(\alpha(x+y))h_N(\alpha(x-y))dy.$$

Setting

$$u = \alpha(y - ip/\alpha^2\hbar) \quad \text{and} \quad \beta = ip/\alpha\hbar,$$

we get, after simplification and using the parity relation (9.22),

$$W\Phi_N(z) = K_N(z) \int_{-\infty}^{\infty} e^{-u^2} h_N(u + \alpha x + \beta) h_N(u - \alpha x + \beta) du, \quad (9.32)$$

where

$$K_N(z) = C_N \frac{(-1)^N}{\alpha} e^{-(\alpha^2 x^2 - \beta^2)} = \frac{(-1)^N}{\pi \hbar} \frac{\pi^{-1/2}}{2^N N!} e^{-\frac{2}{\hbar\omega} H(z)}.$$

Using formula (9.30) one finds that

$$\int_{-\infty}^{\infty} e^{-u^2} h_N(u + \alpha x + \beta) h_N(u - \alpha x + \beta) du = 2^N \sqrt{\pi} N! L_N \left( \frac{4}{\hbar\omega} H(z) \right),$$

where $L_N$ is the Laguerre polynomial (9.28), hence, after some simplifications,

$$W\Phi_N(z) = \frac{(-1)^N}{\pi \hbar} e^{-\frac{2}{\hbar\omega} H(z)} L_N \left( \frac{4}{\hbar\omega} H(z) \right)$$

which is formula (9.31). □

**Example 124.** Assume that $m = \omega = 1$ so that $H = \frac{1}{2}(p^2 + x^2)$. Then (9.31) becomes

$$W\Phi_N(x, p) = \frac{(-1)^N}{\pi \hbar} e^{-\frac{1}{\hbar}|z|^2} L_N \left( \frac{2}{\hbar}|z|^2 \right). \quad (9.33)$$

Assume that $m = \omega = 1$ and $\hbar = 1$; then $\Phi_N$ is the standard Hermite function $H_N$ and

$$W H_N(x, p) = \frac{(-1)^N}{\pi} e^{-|z|^2} L_N \left( 2|z|^2 \right). \quad (9.34)$$

Taking in addition $\hbar = 1/2\pi$ we get

$$W\Phi_N(x, p) = 2(-1)^N e^{-2\pi|z|^2} L_N \left( 4\pi|z|^2 \right). \quad (9.35)$$

### 9.2.3. The cross-Wigner transform of Hermite functions

Less known is the cross-Wigner transform of Hermite functions (see however [1, 82]). We give a simple proof here using the Gradshteyn and Ryzhik formula (9.30).

**Proposition 125.** *Let $M \leq N$. The cross-Wigner transform of $\Phi_M$ and $\Phi_N$ is given by*

$$W(\Phi_M, \Phi_N)(z) = C_{M,N}\left(x - \frac{ip}{m\omega}\right)^{N-M} e^{-\frac{2}{\hbar\omega}H(z)} L_M^{(N-M)}\left(\frac{4}{\hbar\omega}H(z)\right),$$

$$(9.36)$$

*where the constant factor $C_{M,N}$ is given by*

$$C_{M,N} = \frac{(-1)^M}{\pi\hbar}\sqrt{\frac{2^N M!}{2^M N!}}\left(\frac{m\omega}{\hbar}\right)^{(N-M)/2}.$$

**Proof.** In the course of the proof of Proposition 123, we showed that

$$W\Phi_N(z) = K_N(z)\int_{-\infty}^{\infty} e^{-u^2} h_N(u + \alpha x + \beta) h_N(u - \alpha x + \beta)du, \quad (9.37)$$

where

$$K_N(z) = \frac{(-1)^N}{\pi\hbar}\frac{\pi^{-1/2}}{2^N N!}e^{-\frac{2}{\hbar\omega}H(z)}.$$

In the present case, a similar argument leads to the formula

$$W(\Phi_M, \Phi_N)(z) = K_{M,N}(z)I_{M,N}(z),$$

where the factor $K_{M,N}(z)$ is given by

$$K_{M,N}(z) = \frac{(-1)^N}{\pi\hbar}\sqrt{\frac{1}{2^{M+N}M!N!}}\pi^{-1/2}e^{-\frac{2}{\hbar\omega}H(z)}$$

and

$$I_{M,N}(z) = \int_{-\infty}^{\infty} e^{-u^2} h_M(u + \alpha x + \beta) h_N(u - \alpha x + \beta)du.$$

In view of formula (9.30) we have

$$I_{M,N}(z) = 2^N\sqrt{\pi}M!(-1)^{N-M}(\alpha x - \beta)^{N-M} L_M^{(N-M)}(2(\alpha^2 x^2 - \beta^2));$$

using the relations

$$\alpha x - \beta = \sqrt{\frac{m\omega}{\hbar}}\left(x - \frac{ip}{m\omega}\right), \quad 2(\alpha^2 x^2 - \beta^2) = \frac{4}{\hbar\omega}H(z),$$

this is

$$I_{M,N}(z) = 2^N M! (-1)^{N-M} \left(\frac{m\omega}{\hbar}\right)^{(N-M)/2}$$

$$\times \left(x - \frac{ip}{m\omega}\right)^{N-M} L_M^{(N-M)} \left(\frac{4}{\hbar\omega} H(z)\right)$$

hence formula (9.36) after a few simplifications. □

**Example 126.** Choose $m = \omega = 1$ and $\hbar = 1/2\pi$. Introducing the complex variable $\zeta = x + ip$ formula (9.36) then reads

$$W(\Phi_M, \Phi_N)(z) = C_{M,N} \bar{\zeta}^{N-M} e^{-2\pi|\zeta|^2} L_M^{(N-M)} \left(4\pi|\zeta|^2\right) \qquad (9.38)$$

with

$$C_{M,N} = 2^{N-M+1}(-1)^M \pi^{(N-M)/2} \sqrt{\frac{M!}{N!}}$$

(cf. the formula in Theorem 1.105 in [28]).

The cross-ambiguity function of $(\Phi_M, \Phi_N)$ is easily computed.

**Corollary 127.** *Let $M \leq N$. We have*

$$A(\Phi_M, \Phi_N)(z) = D_{M,N} \left(x - \frac{ip}{m\omega}\right)^{N-M} e^{-\frac{1}{2\hbar\omega} H(z)} L_M^{(N-M)} \left(\frac{1}{\hbar\omega} H(z)\right),$$
$$(9.39)$$

*where*

$$D_{M,N} = \frac{(-1)^{M+N}}{2\pi\hbar} 2^{N-M} \sqrt{\frac{2^N M!}{2^M N!}} \left(\frac{m\omega}{\hbar}\right)^{(N-M)/2}.$$

*In particular*

$$A\Phi_N(z) = \frac{1}{2\pi\hbar} e^{-\frac{1}{2\hbar\omega} H(z)} L_N \left(\frac{1}{\hbar\omega} H(z)\right). \qquad (9.40)$$

**Proof.** The relation (3.17) in Chapter 3 becomes here:

$$A(\psi, \phi)(z) = \frac{1}{2} W(\psi, \phi^\vee) \left(\frac{1}{2} z\right). \qquad (9.41)$$

Using the parity formula (9.22) we have $\Phi_N^\vee(x) = (-1)^N \Phi_N(x)$ hence

$$A(\Phi_M, \Phi_N)(z) = \frac{1}{2}(-1)^N W(\Phi_M, \Phi_N) \left(\frac{1}{2} z\right).$$

□

**Example 128.** Choose $m = \omega = 1$ and $\hbar = 1/2\pi$. Formula (9.40) becomes

$$A\Phi_N(z) = e^{-\frac{\pi}{2}(x^2+p^2)} L_N\left(\pi(x^2 + p^2)\right).$$

### 9.2.4. Flandrin's conjecture

Flandrin has conjectured that

$$\int_\Omega W\psi(z)dz \leq ||\psi||$$

for all $\psi \in L^2(\mathbb{R}^n)$ when $\Omega$ is a convex subset of the phase plane $\mathbb{R}^2$. The conjecture of Flandrin says that over convex sets $\Omega$ the positive and negative values of the Wigner distribution $W\psi$ cancel each other to an extent that the integral over $\Omega$ does not exceed the integral over the whole plane. In particular the set of $z$ such that $W\psi \geq 0$ can never be a convex set, unless it is the whole plane. Flandrin [27] proved in 1986 that this conjecture is true for balls of radius $\sqrt{\hbar}$, and Janssen showed in [61] that this follows from the fact that the Wigner distribution of Hermite functions are Laguerre polynomials. Flandrin's result has been generalized to the $n$-dimensional case by Lieb and Ostrover [64]. They actually address the following Wigner distribution localization problem: given a measurable set $D \subset \mathbb{R}^{2n}$ find the best possible bounds to the localization function

$$L(D) = \sup_\psi \int_D W\psi(z)dz,$$

where the supremum is taken over all the functions $\psi \in L^2(\mathbb{R}^n)$ such that $||\psi||_{L^2} = 1$. Lieb and Ostrover then show that the standard Gaussian $\phi_0(x) = (\pi\hbar)^{-n/4} e^{-|x|^2/2\hbar}$ is the unique maximizer of the Wigner distribution localization problem when $D$ is a ball $B^{2n}(r)$ in $\mathbb{R}^{2n}$ centered at the origin. In particular,

$$L(B^{2n}(r)) = 1 - \frac{\Gamma(n, r^2)}{(n-1)^2},$$

where

$$\Gamma(s, x) = \int_x^\infty e^{-t} t^{s-1} dt$$

is the upper incomplete gamma function.

## Main Formulas in Chapter 9

Gaussian $\qquad\qquad\qquad \psi_M(x) = \left(\frac{1}{\pi\hbar}\right)^{n/4} (\det X)^{1/4} e^{-\frac{1}{2\hbar}Mx^2}$

Wigner transform $\qquad W\psi_M(z) = \left(\frac{1}{\pi\hbar}\right)^n e^{-\frac{1}{\hbar}Gz^2}$

$$G = \begin{pmatrix} X + YX^{-1}Y & YX^{-1} \\ X^{-1}Y & X^{-1} \end{pmatrix}$$

Ambiguity function $\qquad A\psi_M(z) = \left(\frac{1}{2\pi\hbar}\right)^n e^{-\frac{1}{4\hbar}Gz^2}$

Hermite polynomial $\qquad h_N(x) = (-1)^N e^{x^2}\left(\frac{d^N}{dx^N}e^{-x^2}\right)$

Hermite function $\qquad H_N(x) = \sqrt{\frac{1}{2^N N!}}\left(\frac{1}{\pi}\right)^{1/4} e^{-x^2/2}h_N(x)$

Laguerre polynomial $\qquad L_N(x) = e^x \frac{d^N}{dx^N}(x^N e^{-x})$

Wigner $(\hbar = 1)$ $\qquad WH_N(x,p) = \frac{(-1)^N}{\pi}e^{-|z|^2}L_N\left(2|z|^2\right)$

# Chapter 10

# Sub-Gaussian Estimates

**Summary 129.** Hardy's theorem on Fourier transforms implies that the Wigner transform cannot be arbitrarily sharply concentrated peaked around a phase space point.

In this chapter we study a multidimensional version of Hardy's uncertainty principle and its implication for Wigner transforms dominated by a phase space Gaussian. We will give two different proofs of this result.

## 10.1. Hardy's Uncertainty Principle

There are numerous versions of the uncertainty principle in the literature. Hardy's uncertainty principle is in a sense archetypical since it clearly shows the essential role played by the Fourier transform in these questions.

### 10.1.1. The one-dimensional case

A function $\psi \in L^2(\mathbb{R}^n)$ and its Fourier transform

$$F\psi(x) = \left(\frac{1}{2\pi\hbar}\right)^{n/2} \int_{\mathbb{R}^n} e^{-\frac{i}{\hbar}x\cdot x'}\psi(x')dx'$$

cannot be simultaneously arbitrarily sharply localized. Assume for instance that $\psi$ is of compact support; then, by the Paley–Wiener theorem its Fourier transform can be extended to an entire function on the complex plane, and is hence never of compact support. A precise way to express this trade-off between a function and its Fourier transform was discovered in 1933 by Hardy [57]. He showed, using the Phragmén–Lindelöf principle from complex analysis, that if a function $\psi \in L^2(\mathbb{R})$ and its Fourier transform

$$F\psi(x) = \left(\frac{1}{2\pi\hbar}\right)^{1/2} \int_{-\infty}^{\infty} e^{-\frac{i}{\hbar}xx'}\psi(x')dx'$$

simultaneously satisfy the estimates

$$|\psi(x)| \le Ce^{-\frac{a}{2\hbar}x^2}, \quad |F\psi(p)| \le Ce^{-\frac{b}{2\hbar}p^2} \tag{10.1}$$

for some $a, b > 0$ and $C > 0$, then we have the following conditions:

- If $ab > 1$, then $\psi = 0$.
- If $ab = 1$, we have $\psi(x) = Ce^{-\frac{a}{2\hbar}x^2}$ for some complex constant $C$.
- If $ab < 1$, then $\psi$ is a function of the type $\psi(x) = Q(x)e^{-\frac{a}{2\hbar}x^2}$ where $Q$ is a polynomial.

We are going to prove a multidimensional version of Hardy's theorem. For this we will need the two lemmas below.

### 10.1.2. Two lemmas

The first lemma is a diagonalization result (see Appendix C for a proof).

**Lemma 130.** *Let $A$ and $B$ be two positive-definite symmetric $n \times n$ matrices. There exists $L \in \mathrm{GL}(n, \mathbb{R})$ such that*

$$L^T AL = L^{-1}B(L^T)^{-1} = \Lambda, \tag{10.2}$$

*where $\Lambda$ is the diagonal matrix*

$$\Lambda = \begin{pmatrix} \sqrt{\lambda_1} & 0 & \cdots & 0 \\ 0 & \sqrt{\lambda_2} & \cdots & 0 \\ \vdots & \vdots & \ddots & \vdots \\ 0 & 0 & \cdots & \sqrt{\lambda_n} \end{pmatrix} \tag{10.3}$$

*the positive numbers $\lambda_1, \ldots, \lambda_n$ being the eigenvalues of $AB$.*

Notice that the eigenvalues of $AB$ are real since they are the same as those of the symmetric positive definite matrix $A^{1/2}BA^{1/2}$.

We will also need the following elementary result about tensor products of functions.

**Lemma 131.** *Let $n > 1$. For $1 \leq j \leq n$ let $f_j$ be a function of $(x_1, \ldots, \widetilde{x}_j, \ldots, x_n) \in \mathbb{R}^{n-1}$ (the tilde~ suppressing the term it covers), and $g_j$ a function of $x_j \in \mathbb{R}$. If*

$$h = f_1 \otimes g_1 = \cdots = f_n \otimes g_n, \tag{10.4}$$

*then there exists a constant $C$ such that $h = Cg_1 \otimes \cdots \otimes g_n$.*

**Proof.** Let us first prove the result for $n = 2$. In this case (10.4) is

$$h(x_1, x_2) = f_1(x_2)g_1(x_1) = f_2(x_1)g_2(x_2).$$

If $g_1(x_1)g_2(x_2) \neq 0$ then

$$\frac{f_1(x_2)}{g_2(x_2)} = \frac{f_2(x_1)}{g_1(x_1)} = C$$

hence $f_1(x_2) = Cg_2(x_2)$ and $h(x_1, x_2) = Cg_1(x_1)g_2(x_2)$. If $g_1(x_1)g_2(x_2) = 0$ then $h(x_1, x_2) = 0$ hence $h(x_1, x_2) = Cg_1(x_1)g_2(x_2)$ in all cases. The general case follows by induction on the dimension $n$: suppose that

$$h = f_1 \otimes g_1 = \cdots = f_n \otimes g_n = f_{n+1} \otimes g_{n+1};$$

for fixed $x_{n+1}$ the function

$$k = f_1 \otimes g_1 = \cdots = f_n \otimes g_n$$

is given by

$$k(x, x_{n+1}) = C(x_{n+1})g_1(x_1) \cdots g_n(x_n),$$

where $C(x_{n+1})$ only depends on $x_{n+1}$. Since

$$k(x, x_{n+1}) = f_{n+1}(x_1, \ldots, x_n)g_{n+1}(x_{n+1}),$$

it follows that $C(x_{n+1})$ is a constant $C$. $\qquad\square$

### 10.1.3. The multidimensional Hardy uncertainty principle

Let us generalize Hardy's theorem on Fourier transforms to the case of several variables. We will be following closely [39, 44, 46].

**Proposition 132.** *Let $A$ and $B$ be two real positive definite matrices and $\psi \in L^2(\mathbb{R}^n)$, $\psi \neq 0$. Assume that*

$$|\psi(x)| \leq Ce^{-\frac{1}{2\hbar}Ax^2} \quad \text{and} \quad |F\psi(p)| \leq Ce^{-\frac{1}{2\hbar}Bp^2} \tag{10.5}$$

*for some constant $C > 0$. Then we have the following assertions.*

(i) *The eigenvalues $\lambda_j, j = 1, \ldots, n$, of $AB$ are $\leq 1$.*
(ii) *If $\lambda_j = 1$ for all $j$, then $\psi(x) = Ce^{-\frac{1}{2\hbar}Ax^2}$ for some complex constant $C$.*
(iii) *If $\lambda_j < 1$ for some index $j$, then $\psi(x) = Q(x)e^{-\frac{1}{2\hbar}Ax^2}$ for some polynomial $Q$.*

**Proof.** (i) Let $L$ be as in Lemma 130 and order the eigenvalues of $AB$ decreasingly: $\lambda_1 \geq \lambda_2 \geq \cdots \geq \lambda_n$. It suffices then to show that $\lambda_1 \leq 1$. Setting $\psi_L(x) = \psi(Lx)$ we have

$$F\psi_L(p) = F\psi((L^T)^{-1}p);$$

condition (10.5) is equivalent to

$$|\psi_L(x)| \leq Ce^{-\frac{1}{2\hbar}\Lambda x^2}, \quad |F\psi_L(p)| \leq Ce^{-\frac{1}{2\hbar}\Lambda p^2}, \tag{10.6}$$

where $\Lambda$ is the diagonal matrix (C.4). Setting $\psi_{L,1}(x_1) = \psi_L(x_1, 0, \ldots, 0)$ we have

$$|\psi_{L,1}(x_1)| \leq Ce^{-\frac{1}{2\hbar}\lambda_1 x_1^2}. \tag{10.7}$$

On the other hand, by the Fourier inversion formula,

$$\int_{\mathbb{R}^{n-1}} F\psi_L(p)dp_2 \cdots dp_n = (2\pi\hbar)^{n/2} \int_{\mathbb{R}^{2n-1}} e^{-\frac{i}{\hbar}p \cdot x}\psi_L(x)dxdp_2 \cdots dp_n$$

$$= (2\pi\hbar)^{(n-1)/2}F\psi_{L,1}(p_1)$$

and hence

$$|F\psi_{L,1}(p_1)| \leq C_{L,1}e^{-\frac{1}{2\hbar}\lambda_1 p_1^2} \tag{10.8}$$

for some constant $C_{L,1} > 0$. Applying Hardy's theorem to the inequalities (10.7) and (10.8), we must have $\lambda_1^2 \leq 1$ hence the assertion(i).

(ii) The condition $\lambda_j = 1$ sfor all $j$ means that

$$|\psi_L(x)| \leq Ce^{-\frac{1}{2\hbar}x^2}, \quad |F\psi_L(p)| \leq Ce^{-\frac{1}{2\hbar}p^2} \tag{10.9}$$

for some $C > 0$. Let us keep $x' = (x_2, \ldots, x_n)$ constant; the partial Fourier transform of $\psi_L$ in the $x_1$ variable is $F_1\psi_L = (F')^{-1}F\psi_L$ where $(F')^{-1}$ is the inverse Fourier transform in the $x'$ variables, hence there exists $C' > 0$ such that

$$|F_1\psi_L(x_1, x')| \leq \left(\frac{1}{2\pi\hbar}\right)^{\frac{n-1}{2}} \int_{\mathbb{R}^{n-1}} |F\psi_L(p)|dp_2\cdots dp_n \leq C'e^{-\frac{1}{2\hbar}p_1^2}.$$

Since $|\psi_L(x)| \leq C(x')e^{-\frac{1}{2\hbar}x_1^2}$ with $C(x') \leq e^{-\frac{1}{2\hbar}x'^2}$, it follows from Hardy's theorem in the case $n = 1$ that we can write

$$\psi_L(x) = f_1(x')e^{-\frac{1}{2\hbar}x_1^2}$$

for some real $C^\infty$ function $f_1$ on $\mathbb{R}^{n-1}$. Applying the same argument to the remaining variables $x_2, \ldots, x_n$ we conclude that there exist $C^\infty$ functions $f_j$ for $j = 2, \ldots, n$, such that

$$\psi_L(x) = f_j(x_1, \ldots, \widetilde{x}_j, \ldots, x_n)e^{-\frac{1}{2\hbar}x_1^2}. \tag{10.10}$$

In view of Lemma 131, we have $\psi_L(x) = C_L e^{-\frac{1}{2\hbar}x^2}$ for some constant $C_L$; since $\Lambda = I = L^T AL$ we thus have $\psi(x) = C_L e^{-Ax^2/2\hbar}$ as claimed.

(iii) Assume that $\lambda_1 < 1$ for $j \in \mathcal{J} \subset \{1, \ldots, n\}$. By the same argument as in the proof of part (ii) where we established formula (10.10), we infer, using Hardy's theorem in the case $ab < 1$, that each function

$$\psi_L(x) = f_j(x_1, \ldots, \widetilde{x}_j, \ldots, x_n)Q_j(x_j)e^{-\frac{1}{2\hbar}x_j^2}$$

($Q_j$ a polynomial) satisfies the estimates (10.5) for some constants $C_A, C_B > 0$. One concludes the proof by using once again Lemma 131. $\square$

## 10.2. Sub-Gaussian Estimates for the Wigner Transform

We now apply the Hardy uncertainty principle to study inequalities of the type

$$W\psi(z) \leq Ce^{-\frac{1}{\hbar}Mz^2},$$

where $M$ is a symmetric positive definite matrix. We will call such equalities "sub-Gaussian estimates".

### 10.2.1. Statement of the result

Let $\psi_M$ be the generalized normalized Gaussian considered in Chapter 9:

$$\psi_M(x) = \left(\frac{1}{\pi\hbar}\right)^{n/4}(\det X)^{1/4}e^{-\frac{1}{2\hbar}(X+iY)x^2}, \tag{10.11}$$

where $X$ and $Y$ are real symmetric $n \times n$ matrices, $X$ positive definite. We proved in Proposition 112 that the Wigner transform of $\psi_M$ is the phase space Gaussian

$$W\psi_M(z) = \left(\frac{1}{\pi\hbar}\right)^n e^{-\frac{1}{\hbar}Gz\cdot z}, \tag{10.12}$$

where $G = S^T S$ and

$$S = \begin{pmatrix} X^{1/2} & 0 \\ X^{-1/2}Y & X^{-1/2} \end{pmatrix} \tag{10.13}$$

Explicitly

$$G = \begin{pmatrix} X + YX^{-1}Y & YX^{-1} \\ X^{-1}Y & X^{-1} \end{pmatrix}. \tag{10.14}$$

We now ask whether there are function (or distributions) for which one has better results in terms of phase space "concentration". The answer is as follows.

**Proposition 133.** *Let $\psi \in L^2(\mathbb{R}^n)$, $\psi \neq 0$, and assume that there exists $C > 0$ such that $W\psi(z) \leq Ce^{-\frac{1}{\hbar}Mz^2}$ where $M$ is a positive definite real matrix. Then the symplectic eigenvalues $\lambda_1 \geq \lambda_2 \geq \cdots \geq \lambda_n$ of $M$ are all $\leq 1$. When $\lambda_1 = \lambda_2 = \cdots = \lambda_n = 1$, then $\psi$ is a generalized Gaussian.*

Recall that the $\lambda_j$ are the moduli of the eigenvalues $\pm i\lambda$, $\lambda > 0$, of $JM$. It follows from this result that a Wigner transform $W\psi$ can never have compact support: assume that there exists $R > 0$ such that $W\psi(z) = 0$, for $|z| > R$. Then, for every $a > 0$, there exists a constant $C > 0$ such that

$$W\psi(z) \leq Ce^{-\frac{a}{\hbar}|z|^2} \quad \text{for every } z \in \mathbb{R}^{2n};$$

choosing $a$ large enough this contradicts the conclusion of the proposition.

We are going to give two proofs of Proposition 133; the first is a direct application of the multidimensional Hardy uncertainty principle (Proposition 132), the second uses only the one-dimensional Hardy uncertainty principle, but requires the full power of Williamson's symplectic diagonalization theorem.

### 10.2.2. First proof

Let $L$ be as in Lemma 130. It suffices to show that $\lambda_1 \leq 1$. Setting $\psi_L(x) = \psi(Lx)$, we have

$$F\psi_L(p) = F\psi((L^T)^{-1}p),$$

which is equivalent to

$$|\psi_L(x)| \le Ce^{-\frac{1}{2\hbar}\Lambda x^2} \quad \text{and} \quad |F\psi_L(p)| \le Ce^{-\frac{1}{2\hbar}\Lambda p^2}, \tag{10.15}$$

where the diagonal matrix is given by

$$\Lambda = \begin{pmatrix} \lambda_1 & 0 & \cdots & 0 \\ 0 & \lambda_2 & \cdots & 0 \\ \vdots & \vdots & \ddots & \vdots \\ 0 & 0 & \cdots & \lambda_n \end{pmatrix}.$$

Setting $\psi_{L,1}(x_1) = \psi_L(x_1, 0, \ldots, 0)$ we have

$$|\psi_{L,1}(x_1)| \le Ce^{-\frac{1}{2\hbar}\lambda_1 x_1^2}. \tag{10.16}$$

On the other hand, by the Fourier inversion formula,

$$\int F\psi_L(p)dp_2\cdots dp_n = (2\pi\hbar)^{n/2}\int_{\mathbb{R}\times\mathbb{R}^{n-1}} e^{-\frac{i}{\hbar}p\cdot x}\psi_L(x)dxdp_2\cdots dp_n$$

$$= (2\pi\hbar)^{(n-1)/2}F\psi_{L,1}(p_1)$$

hence we have the inequality

$$|F\psi_{L,1}(p_1)| \le C_{L,1}e^{-\frac{1}{2\hbar}\lambda_1 p_1^2} \tag{10.17}$$

for some constant $C_{L,1} > 0$. Applying Hardy's uncertainty principle in one dimension to the inequalities (10.15), we must have $\lambda_1^2 \le 1$ hence the assertion. We next note that the condition $\lambda_1 = \lambda_2 = \cdots = \lambda_n = 1$ means that

$$|\psi_L(x)| \le Ce^{-\frac{1}{2\hbar}x^2} \quad \text{and} \quad |F\psi_L(p)| \le Ce^{-\frac{1}{2\hbar}p^2} \tag{10.18}$$

for some $C > 0$. Let us keep $x' = (x_2, \ldots, x_n)$ constant; the partial Fourier transform of $\psi_L$ in the $x_1$ variable is $F_1\psi_L = (F')^{-1}F\psi_L$ where $(F')^{-1}$ is the inverse Fourier transform in the $x'$ variables, hence there exists $C' > 0$ such that

$$|F_1\psi_L(x_1, x')| \le \left(\frac{1}{2\pi\hbar}\right)^{\frac{n-1}{2}}\int |F\psi_L(p)|dp_2\cdots dp_n \le C'e^{-\frac{1}{2\hbar}p_1^2}.$$

Since $|\psi_L(x)| \le C(x')e^{-\frac{1}{2\hbar}x_1^2}$ with $C(x') \le e^{-\frac{1}{2\hbar}x'^2}$, it follows from Hardy's theorem that we can write

$$\psi_L(x) = f_1(x')e^{-\frac{1}{2\hbar}x_1^2}$$

for some real $C^\infty$ function $f_1$ on $\mathbb{R}^{n-1}$. Applying the same argument to the remaining variables $x_2, \ldots, x_n$, we conclude that there exist $C^\infty$ functions

$f_j$ for $j = 2, \ldots, n$, such that

$$\psi_L(x) = f_j(x_1, \ldots, \widetilde{x}_j, \ldots, x_n)e^{-\frac{1}{2\hbar}x_1^2}. \tag{10.19}$$

In view of Lemma 131, we have $\psi_L(x) = C_L e^{-\frac{1}{2\hbar}x^2}$ for some constant $C_L$; since $\Lambda = I = L^T A L$, we thus have $\psi(x) = C_L e^{-Ax^2/2\hbar}$ as claimed.

### 10.2.3. Second proof

In view of Williamson's symplectic diagonalization theorem (see Appendix C), we can find $S \in Sp(n)$ such that

$$MSz \cdot Sz = \sum_{j=1}^{n} \lambda_j(x_j^2 + p_j^2), \tag{10.20}$$

where $\lambda_1 \geq \lambda_2 \geq \cdots \geq \lambda_n$ the symplectic eigenvalues of $M$. It follows, using (10.20), that the inequality $W\psi(z) \leq Ce^{-\frac{1}{\hbar}Mz \cdot z}$ is equivalent to

$$W\psi(S^{-1}z) \leq C \exp\left(-\frac{1}{\hbar}\sum_{j=1}^{n}\lambda_j(x_j^2 + p_j^2)\right). \tag{10.21}$$

The next step is to observe that in view of the symplectic covariance property (5.9) of the Wigner transform, we have

$$W\psi(S^{-1}z) = W\widehat{S}\psi(z),$$

where $\widehat{S} \in Mp(n)$ is one of the two metaplectic operators with projection $S \in Sp(n)$. Since $\widehat{S}\psi \in L^2(\mathbb{R}^n)$ and the symplectic spectrum $(\lambda_1, \lambda_2, \ldots, \lambda_n)$ of $M$ is a symplectic invariant, it is thus no restriction to assume $S = I$ and $\widehat{S} = I$. Integrating the inequality

$$W\psi(z) \leq C \exp\left(-\frac{1}{\hbar}\sum_{j=1}^{n}\lambda_j(x_j^2 + p_j^2)\right)$$

in $x$ and $p$, respectively we get, using the marginal properties of the Wigner transform,

$$|\psi(x)| \leq C_1 \exp\left(-\frac{1}{2\hbar}\sum_{j=1}^{n}\lambda_j x_j^2\right), \tag{10.22}$$

$$|F\psi(p)| \leq C_1 \exp\left(-\frac{1}{2\hbar}\sum_{j=1}^{n}\lambda_j p_j^2\right), \tag{10.23}$$

for some constant $C_1 > 0$. Let us now introduce the following notation. We set $\psi_1(x_1) = \psi(x_1, 0, \ldots, 0)$ and denote by $F_1$ the one-dimensional Fourier transform in the $x_1$ variable. Now, we first note that (10.22) implies that

$$|\psi_1(x_1)| \leq C_1 \exp\left(-\frac{\lambda_1}{2\hbar}x_1^2\right). \tag{10.24}$$

On the other hand, by definition of the Fourier transform $F$,

$$\int_{\mathbb{R}^{n-1}} F\psi(p)dp_2 \cdots dp_n = \left(\frac{1}{2\pi\hbar}\right)^{n/2}$$

$$\times \int_{\mathbb{R}^{n-1}} \left(\int_{\mathbb{R}^n} e^{-\frac{i}{\hbar}p\cdot x}\psi(x)dx\right) dp_2 \cdots dp_n;$$

taking into account the Fourier inversion formula this formula can be rewritten as

$$\int_{\mathbb{R}^{n-1}} F\psi(p)dp_2 \cdots dp_n = (2\pi\hbar)^{(n-1)/2} F_1\psi_1(p_1).$$

It follows that

$$|F_1\psi_1(p_1)| \leq \left(\frac{1}{2\pi\hbar}\right)^{(n-1)/2} C_1 \int_{\mathbb{R}^{n-1}} \exp\left(-\frac{1}{2\hbar}\sum_{j=1}^{n}\lambda_j p_j^2\right) dp_2 \cdots dp_n,$$

that is

$$|F_1\psi_1(p_1)| \leq C_3 \exp\left(-\frac{\lambda_1}{2\hbar}p_1^2\right) \tag{10.25}$$

for some constant $C_3 > 0$. Applying Hardy's uncertainty principle we see that the condition $\lambda_j^2 \leq 1$ are both necessary and sufficient for these inequalities to hold (remember that we are using the ordering convention $\lambda_1 \geq \lambda_2 \geq \cdots \geq \lambda_n$). The case $\lambda_1 = \lambda_2 = \cdots = \lambda_n = 1$ is treated as in the first proof given above.

## Main Formulas in Chapter 10

Hardy's UP
$$\left.\begin{array}{l} |\psi(x)| \leq Ce^{-\frac{a}{2\hbar}x^2} \\ |F\psi(p)| \leq Ce^{-\frac{b}{2\hbar}p^2} \end{array}\right\} \Longrightarrow ab \leq 1$$

Hardy's UP (multidimensional)

$$\left.\begin{array}{l} |\psi(x)| \leq Ce^{-\frac{1}{2\hbar}Ax^2} \\ |F\psi(p)| \leq Ce^{-\frac{1}{2\hbar}Bp^2} \end{array}\right\}$$

$$\implies \begin{cases} \text{Eigenvalues} \\ \text{of } AB \leq 1 \end{cases}$$

Sub-Gaussian estimate

$$W\psi(z) \leq Ce^{-\frac{1}{\hbar}Mz^2}, \, M = M^T > 0$$

$$\iff$$

condition

symplectic eigenvalues of $M$ are $\leq 1$

# Part II
# Applications to Quantum Mechanics

Chapter 11

# Moyal Star Product and Twisted Convolution

**Summary 134.** The Moyal star product is an important topic in phase space quantization; it is closely related to the Schrödinger equation in phase space via the notion of Bopp operator, the latter being variants of deformation quantization in Euclidean space.

We are going to study two topics which are easily interpreted in terms of the product of Weyl operators: the Moyal product and its dual, twisted convolution. The Moyal product leads in a natural way to a pseudodifferential calculus on phase space (Bopp calculus) which is, ultimately, a way of doing deformation quantization on $\mathbb{R}^{2n}$. An essential role in the study of these topics is played by the notion of wavepacket transform defined in Chapter 6 using the cross-Wigner transform; they play the role of intertwiners between Weyl calculus and Bopp calculus.

## 11.1. The Moyal Product of Two Symbols

### 11.1.1. Definition of the Moyal product

Recall that we denote by $\widehat{A} = \mathrm{Op_W}(a)$ the Weyl operator with symbol $a$. The definition below is *not* that usually given in the physical literature.

**Definition 135.** Let $a, b \in \mathcal{S}(\mathbb{R}^{2n})$. The Moyal star product (or simply Moyal product) $a \star_\hbar b$ is defined by

$$\mathrm{Op_W}(a \star_\hbar b) = \mathrm{Op_W}(a)\mathrm{Op_W}(b).$$

It is thus the Weyl symbol $c$ of the product operator $\widehat{C} = \widehat{A}\widehat{B}(\widehat{A} = \mathrm{Op_W}(a)$ and $\widehat{B} = \mathrm{Op_W}(b))$.

Note that both $a \star_\hbar b$ and $b \star_\hbar a$ are well-defined when $a$ and $b$ are in $\mathcal{S}(\mathbb{R}^{2n})$ because the operators $\widehat{A}$ and $\widehat{B}$ map in this case $\mathcal{S}(\mathbb{R}^n)$ into itself and can thus be composed in any order. Notice that in general $a \star_\hbar b \neq b \star_\hbar a$: the Moyal product is not commutative. It is however associative since the operator product is (when defined):

$$(a \star_\hbar b) \star_\hbar c = a \star_\hbar (b \star_\hbar c).$$

Let us give two explicit formulas for the Moyal product.

**Proposition 136.** *The Moyal product of $a, b \in \mathcal{S}(\mathbb{R}^{2n})$ is given by the two equivalent formulas:*

$$a \star_\hbar b(z) = \left(\frac{1}{4\pi\hbar}\right)^{2n} \int_{\mathbb{R}^{4n}} e^{\frac{i}{2\hbar}\sigma(z',z'')} a\left(z + \frac{1}{2}z'\right) b\left(z - \frac{1}{2}z''\right) dz'dz'' \tag{11.1}$$

*and*

$$a \star_\hbar b(z) = \left(\frac{1}{\pi\hbar}\right)^{2n} \int_{\mathbb{R}^{4n}} e^{-\frac{2i}{\hbar}\sigma(u-z,v-z)} a(u)b(v)dudv. \tag{11.2}$$

**Proof.** Let $K$ be the distributional kernel of the product $\widehat{C} = \widehat{A}\widehat{B}$. We have

$$K(x,y) = \left(\frac{1}{2\pi\hbar}\right)^{2n} \int_{\mathbb{R}^{3n}} e^{\frac{i}{\hbar}((x-\alpha)\cdot p+(\alpha-y)\cdot p)} a\left(\frac{1}{2}(x+\alpha),\zeta\right)$$
$$\times b\left(\frac{1}{2}(x+y),\xi\right) d\alpha d\zeta d\xi.$$

In view of formula (4.14) in the previous chapter, the Weyl symbol $c$ of $\widehat{C}$ is given by

$$c(x,p) = \int e^{-\frac{i}{\hbar}p\cdot u} K\left(x + \frac{1}{2}u, x - \frac{1}{2}u\right) du$$

and the Weyl symbol of $\widehat{A}\widehat{B}$ is thus

$$c(z) = \left(\frac{1}{\pi\hbar}\right)^{2n} \int_{\mathbb{R}^{4n}} e^{\frac{i}{\hbar}\Phi} a\left(\frac{1}{2}\left(x + \alpha + \frac{1}{2}u\right), \zeta\right)$$

$$\times b\left(\frac{1}{2}\left(x + \alpha - \frac{1}{2}u\right), \xi\right) d\alpha d\zeta du d\xi,$$

where the phase $\Phi$ is given by

$$\Phi = \left(x - \alpha + \frac{1}{2}u\right)\zeta + \left(\alpha - x + \frac{1}{2}u\right)\xi - u \cdot p$$

$$= \left(x - \alpha + \frac{1}{2}u\right)(\zeta - p) + \left(\alpha - x + \frac{1}{2}u\right)(\xi - p).$$

Introducing the new variables $\zeta' = \zeta - p$, $\xi' = \xi - p$, $\alpha' = \frac{1}{2}(\alpha - x + \frac{1}{2}u)$ and $u' = \frac{1}{2}(\alpha - x - \frac{1}{2}u)$, we have

$$d\alpha d\zeta du d\xi = 2^{2n} d\alpha' \zeta' du' d\xi'$$

and $\Phi = 2\sigma(u', \xi'; \alpha', \zeta')$, hence

$$c(z) = \left(\frac{1}{\pi\hbar}\right)^{2n} \int_{\mathbb{R}^{4n}} e^{\frac{2i}{\hbar}\sigma(u',\xi';\alpha',\zeta')} a(x + \alpha', p + \zeta')$$

$$\times b(x + u', p + \xi') d\alpha' d\zeta' du' d\xi';$$

formula (11.1) follows setting $z' = 2(\alpha', \zeta')$ and $z'' = -2(u', \xi')$. To prove formula (11.2) it suffices to make the change of variables $u = z + \frac{1}{2}z'$, $v = z - \frac{1}{2}z''$ in (11.1). $\quad\square$

**Remark 137.** The exponent in the integral in the right-hand side of (11.2) can be written

$$\sigma(u - z, v - z) = \sigma(z, u) - \sigma(z, v) + \sigma(u, v).$$

Viewing $\sigma$ as a 1-chain $\sigma(u - z, v - z)$ is thus the coboundary $\partial\sigma(z, u, v)$ of $\sigma$, and we can rewrite (11.2) as

$$c(z) = \left(\frac{1}{\pi\hbar}\right)^{2n} \int_{\mathbb{R}^{4n}} e^{-\frac{2i}{\hbar}\partial\sigma(z,u,v)} a(u)b(v) du dv. \qquad (11.3)$$

Formula (11.2) in Proposition 136 shows that the Moyal product $a \star_\hbar b$ is defined when $a, b \in L^1(\mathbb{R}^{2n})$. In fact

$$|a \star_\hbar b(z)| \leq \left( \frac{1}{\pi \hbar} \right)^{2n} \int_{\mathbb{R}^{4n}} |a(u)b(v)| du dv$$

$$= \left( \frac{1}{\pi \hbar} \right)^{2n} \int_{\mathbb{R}^{2n}} |a(u)| du \int_{\mathbb{R}^{2n}} |b(v)| dv.$$

The star product is often "defined" in physics by Groenewold's formula

$$a \star_\hbar b = a \exp \left( \frac{i\hbar}{2} \left[ \overleftarrow{\partial_x} \cdot \overrightarrow{\partial_p} - \overleftarrow{\partial_p} \cdot \overrightarrow{\partial_x} \right] \right) b; \qquad (11.4)$$

the exponential in the right-hand side (the "Janus operator") is understood as a power series, the arrows indicating the direction in which the derivatives act. We leave it to the reader to check that this power series coincides formally with the correct definition above. We will not use Groenewold's formula in this book.

### 11.1.2. Twisted convolution

Closely related to the Moyal product is the twisted convolution of two phase space functions.

**Definition 138.** Let $a, b \in \mathcal{S}(\mathbb{R}^{2n})$. The twisted convolution $a \# b$ is defined by

$$a \# b = F_\sigma(F_\sigma a \star_\hbar F_\sigma b)$$

or, equivalently, by

$$F_\sigma(a \# b) = F_\sigma a \star_\hbar F_\sigma b$$

where $F_\sigma$ is the symplectic Fourier transform.

The equivalence of both definitions is due to the fact that the symplectic Fourier transform is involutive (Appendix B). The twisted convolution of two symbols $a$ and $b$ is thus the twisted Weyl symbol $c_\sigma$ of the product $\widehat{C} = \widehat{A}\widehat{B}$ where $\widehat{A} = \mathrm{Op_W}(a)$ and $\widehat{B} = \mathrm{Op_W}(b)$.

Here is an explicit formula for the twisted convolution.

**Proposition 139.** *We have*

$$a \# b(z) = \left(\frac{1}{2\pi\hbar}\right)^n \int_{\mathbb{R}^{2n}} e^{\frac{i}{2\hbar}\sigma(z,z')} a_\sigma(z - z') b_\sigma(z') dz'; \tag{11.5}$$

*equivalently, the twisted symbol $c_\sigma$ of the product $\widehat{C} = \widehat{A}\widehat{B}$ is given by*

$$c_\sigma(z) = \left(\frac{1}{2\pi\hbar}\right)^n \int_{\mathbb{R}^{2n}} e^{\frac{i}{2\hbar}\sigma(z,z')} a_\sigma(z - z') b_\sigma(z') dz'. \tag{11.6}$$

**Proof.** Let us write the operators $\widehat{A}$ and $\widehat{B}$ in the form (4.2) (Definition 37 in Chapter 4):

$$\widehat{A} = \left(\frac{1}{2\pi\hbar}\right)^n \int_{\mathbb{R}^{2n}} a_\sigma(z_0)\widehat{T}(z_0) dz_0,$$

$$\widehat{B} = \left(\frac{1}{2\pi\hbar}\right)^n \int_{\mathbb{R}^{2n}} b_\sigma(z_1)\widehat{T}(z_1) dz_1.$$

We have, using the property

$$\widehat{T}(z_0 + z_1) = e^{-\frac{i}{2\hbar}\sigma(z_0,z_1)}\widehat{T}(z_0)\widehat{T}(z_1)$$

of Heisenberg–Weyl operators,

$$\widehat{T}(z_0)\widehat{B} = \left(\frac{1}{2\pi\hbar}\right)^n \int_{\mathbb{R}^{2n}} b_\sigma(z_1)\widehat{T}(z_0)\widehat{T}(z_1) dz_1$$

$$= \left(\frac{1}{2\pi\hbar}\right)^n \int_{\mathbb{R}^{2n}} e^{\frac{i}{2\hbar}\sigma(z_0,z_1)} b_\sigma(z_1)\widehat{T}(z_0 + z_1) dz_1$$

and hence

$$\widehat{A}\widehat{B} = \left(\frac{1}{2\pi\hbar}\right)^{2n} \int_{\mathbb{R}^{4n}} e^{\frac{i}{2\hbar}\sigma(z_0,z_1)} a_\sigma(z_0) b_\sigma(z_1)\widehat{T}(z_0 + z_1) dz_0 dz_1.$$

Setting $z = z_0 + z_1$ and $z' = z_1$ this formula can be written

$$\widehat{A}\widehat{B} = \left(\frac{1}{2\pi\hbar}\right)^{2n} \int_{\mathbb{R}^{2n}} \left(\int_{\mathbb{R}^{2n}} e^{\frac{i}{2\hbar}\sigma(z,z')} a_\sigma(z - z') b_\sigma(z') dz'\right) \widehat{T}(z) dz$$

hence formula (11.6). $\qquad\qquad\square$

The twisted convolution is defined when the twisted symbols are integrable: formula (11.5) implies that

$$|a\#b(z)| \leq \left(\frac{1}{2\pi\hbar}\right)^n \int_{\mathbb{R}^{2n}} |a_\sigma(z - z')b_\sigma(z')|dz'$$

and hence, using Fubini's theorem

$$|a\#b(z)| \leq \left(\frac{1}{2\pi\hbar}\right)^n ||a_\sigma||'_{L^1(\mathbb{R}^n)}||b_\sigma||_{L^1(\mathbb{R}^n)}.$$

## 11.2. Bopp Operators

The Weyl correspondence transforms symbols in $\mathcal{S}(\mathbb{R}^{2n})$ into trace class operators and symbols in $L^2(\mathbb{R}^{2n})$ into Hilbert–Schmidt operators. This fact, as Maillard [67] notes, leads to a general method of quantization already suggested by Wigner and Moyal and developed by Flato, Lichnerowicz, Fronsdal, Sternheimer and Bayen in [4,5]; this method is known under the name of star-product method or deformation quantization. The theory of Bopp pseudodifferential operators which we have developed in [37,39,47] is an alternative way of doing Bayen deformation quantization on Euclidean spaces.

### 11.2.1. Bopp shifts

Everything in this section stems from the following observation.

**Lemma 140.** *Let* $a \in \mathcal{S}(\mathbb{R}^{2n})$. *We have*

$$x_j \star_\hbar a = \left(x_j + \frac{1}{2}i\hbar\frac{\partial}{\partial p_j}\right) a, \quad p \star_\hbar a = \left(p_j - \frac{1}{2}i\hbar\frac{\partial}{\partial x_j}\right) a. \quad (11.7)$$

**Proof.** Let us prove this directly without appealing to Proposition 136. It is sufficient to assume that $n = 1$. We thus have to show that the Weyl symbol of $\mathrm{Op_W}(x)\mathrm{Op_W}(a)$ is $xa + \frac{1}{2}i\hbar\frac{\partial a}{\partial p}$ and that of $\mathrm{Op_W}(p)\mathrm{Op_W}(a)$ is $pa - \frac{1}{2}i\hbar\frac{\partial a}{\partial x}$. Let us prove the first identity (11.7). Now $\mathrm{Op_W}(x)$ is just multiplication by $x$ so it suffices to determine the Weyl symbol of $x\mathrm{Op_W}(a)$. Using the integral representation (4.10) given in Chapter 4 we have, for $\psi \in \mathcal{S}(\mathbb{R})$

$$x\widehat{A}\psi(x) = \frac{1}{2\pi\hbar} \int_{\mathbb{R}^2} e^{\frac{i}{\hbar}p(x-y)}xa\left(\frac{1}{2}(x+y), p\right) \psi(y)dydp$$

with $\widehat{A} = \mathrm{Op_W}(a)$. Using the elementary equality $x = \frac{1}{2}(x+y) + \frac{1}{2}(x-y)$ this can be rewritten

$$x\widehat{A}\psi(x) = \frac{1}{2\pi\hbar}\int_{\mathbb{R}^2} e^{\frac{i}{\hbar}p(x-y)}\frac{1}{2}(x+y)a\left(\frac{1}{2}(x+y),p\right)\psi(y)dydp$$

$$+\frac{1}{4\pi\hbar}\int_{\mathbb{R}^2} e^{\frac{i}{\hbar}p(x-y)}(x-y)a\left(\frac{1}{2}(x+y),p\right)\psi(y)dydp.$$

The first integral is

$$\frac{1}{2\pi\hbar}\int_{\mathbb{R}^2} e^{\frac{i}{\hbar}p(x-y)}\frac{1}{2}(x+y)a\left(\frac{1}{2}(x+y),p\right)\psi(y)dydp = \mathrm{Op_W}(xa)\psi(x)$$

and the second is

$$\frac{1}{4\pi\hbar}\int_{\mathbb{R}^2} e^{\frac{i}{\hbar}p(x-y)}(x-y)a\left(\frac{1}{2}(x+y),p\right)\psi(y)dydp$$

$$= \frac{1}{4\pi i}\int_{\mathbb{R}^2}\frac{\partial}{\partial p}(e^{\frac{i}{\hbar}p(x-y)})a\left(\frac{1}{2}(x+y),p\right)\psi(y)dydp$$

that is, after an integration by parts in the variable $p$,

$$\frac{1}{4\pi\hbar}\int_{\mathbb{R}^2} e^{\frac{i}{\hbar}p(x-y)}(x-y)a\left(\frac{1}{2}(x+y),p\right)\psi(y)dydp$$

$$= \frac{1}{2}i\hbar\mathrm{Op_W}\left(\frac{\partial a}{\partial p}\right)\psi(x).$$

Summarizing, we get

$$x\widehat{A}\psi(x) = \mathrm{Op_W}(xa)\psi(x) + \frac{1}{2}i\hbar\mathrm{Op_W}\left(\frac{\partial a}{\partial p}\right)\psi(x)$$

proving the first identity (11.7). To prove the second identity we note that $\mathrm{Op_W}(p)$ is just $-i\hbar(d/dx)$ hence $\mathrm{Op_W}(p)\mathrm{Op_W}(a)$ is defined by

$$-i\hbar\frac{d}{dx}\widehat{A}\psi(x) = \frac{1}{2\pi i}\int_{\mathbb{R}^2}\frac{d}{dx}\left(e^{\frac{i}{\hbar}p(x-y)}a\left(\frac{1}{2}(x+y),p\right)\right)\psi(y)dydp.$$

Using Leibniz' rule, the right-hand side is the sum of the integrals

$$\frac{1}{2\pi i}\int_{\mathbb{R}^2}\frac{d}{dx}(e^{\frac{i}{\hbar}p(x-y)})a\left(\frac{1}{2}(x+y),p\right)\psi(y)dydp$$

$$=\frac{1}{2\pi\hbar}\int_{\mathbb{R}^2}e^{\frac{i}{\hbar}p(x-y)}pa\left(\frac{1}{2}(x+y),p\right)\psi(y)dydp,$$

that is

$$\frac{1}{2\pi i}\int_{\mathbb{R}^2}\frac{d}{dx}(e^{\frac{i}{\hbar}p(x-y)})a\left(\frac{1}{2}(x+y),p\right)\psi(y)dydp=\mathrm{Op_W}(pa)\psi(x)$$

and

$$\frac{1}{4\pi i}\int_{\mathbb{R}^2}e^{\frac{i}{\hbar}p(x-y)}\frac{\partial a}{\partial x}\left(\frac{1}{2}(x+y),p\right)\psi(y)dydp=-\frac{1}{2}i\hbar\mathrm{Op_W}\left(\frac{\partial a}{\partial x}\right).$$

Summarizing,

$$-i\hbar\frac{d}{dx}\widehat{A}\psi(x)=\mathrm{Op_W}(pa)\psi(x)-\frac{1}{2}i\hbar\mathrm{Op_W}\left(\frac{\partial a}{\partial x}\right)$$

proving the second formula (11.7).                                    □

**Definition 141.** The operators

$$\widetilde{x}_j=x_j+\frac{1}{2}i\hbar\frac{\partial}{\partial p_j},\quad\widetilde{p}_j=p_j-\frac{1}{2}i\hbar\frac{\partial}{\partial x_j},$$

which can be written more compactly as

$$\widetilde{z}=z+\frac{1}{2}i\hbar J\partial_z,\quad\widetilde{z}=(\widetilde{x},\widetilde{p}),$$

are called Bopp shifts, following [11].

The Bopp shifts are defined on functions or distributions on phase space. The standard operators $\widehat{x}_j$ and $\widehat{p}_j$ satisfy the canonical commutation relations $[\widehat{x}_j,\widehat{p}_j]=i\hbar$; a simple calculation shows that the Bopp shifts satisfy the similar relations

$$[\widetilde{x}_j,\widetilde{p}_j]=i\hbar. \tag{11.8}$$

## 11.2.2.  Definition and justification of Bopp operators

To motivate the definition of Bopp operators, we are going to rewrite the Moyal product in a particular way. Recall (formula (11.2)) that

$$a \star_\hbar b(z) = \left(\frac{1}{\pi\hbar}\right)^{2n} \int_{\mathbb{R}^{4n}} e^{-\frac{2i}{\hbar}\sigma(u-z,v-z)} a(u)b(v)\,du\,dv.$$

Let us make the change of variables $v = z - \frac{1}{2}z_0$ in the integral; this leads to the new expression

$$a \star_\hbar b(z) = \left(\frac{1}{2\pi\hbar}\right)^{2n} \int_{\mathbb{R}^{4n}} e^{\frac{i}{\hbar}\sigma(u-z,z_0)} a(u) b\left(z - \frac{1}{2}z_0\right) du\,dz_0$$

$$= \left(\frac{1}{2\pi\hbar}\right)^{2n} \int_{\mathbb{R}^{2n}} e^{-\frac{i}{\hbar}\sigma(z,z_0)} b\left(z - \frac{1}{2}z_0\right)$$

$$\times \left(\int_{\mathbb{R}^{2n}} e^{\frac{i}{\hbar}\sigma(u,z_0)} a(u)\,du\right) dz_0.$$

Observing that the integral in the variable $u$ is

$$\int_{\mathbb{R}^{2n}} e^{\frac{i}{\hbar}\sigma(u,z_0)} a(u)\,du = \int_{\mathbb{R}^{2n}} e^{-\frac{i}{\hbar}\sigma(u,z_0)} a(u)\,du = (2\pi\hbar)^n F_\sigma a(z_0),$$

where $F_\sigma a = a_\sigma$ is the symplectic Fourier transform of the symbol $a$ the formula above can be written

$$a \star_\hbar b(z) = \left(\frac{1}{2\pi\hbar}\right)^n \int_{\mathbb{R}^{2n}} a_\sigma(z_0) e^{-\frac{i}{\hbar}\sigma(z,z_0)} b\left(z - \frac{1}{2}z_0\right) dz_0.$$

Introducing the notation

$$\widetilde{T}(z_0)b(z) = e^{-\frac{i}{\hbar}\sigma(z,z_0)} b\left(z - \frac{1}{2}z_0\right), \tag{11.9}$$

we arrive at the following formula for the Moyal product:

$$a \star_\hbar b(z) = \left(\frac{1}{2\pi\hbar}\right)^n \int_{\mathbb{R}^{2n}} a_\sigma(z_0)\widetilde{T}(z_0)b(z)\,dz_0. \tag{11.10}$$

The right-hand side is formally very similar to the expression of a Weyl operator in terms of the Heisenberg–Weyl operator as given in formula (4.2) (Definition 37 in Chapter 4), except that the Heisenberg–Weyl operator is replaced with the operator (11.9) which acts on functions defined on phase space $\mathbb{R}^{2n}$ rather than on functions defined on $\mathbb{R}^n$. Changing slightly notation, we give a name to this new kind of operators.

**Definition 142.** Let $a \in S(\mathbb{R}^{2n})$. The operator

$$\widetilde{A} = \mathrm{Op}_\mathrm{B}(a) : S(\mathbb{R}^{2n}) \longrightarrow S(\mathbb{R}^{2n})$$

defined, for $\Psi \in S(\mathbb{R}^{2n})$, by $\widetilde{A}\Psi = a \star_\hbar \Psi$ is called the Bopp operator with symbol $a$.

The Bopp operator $\widetilde{A} = \mathrm{Op}_\mathrm{B}(a)$ is explicitly given by replacing $b$ with $\Psi \in S(\mathbb{R}^{2n})$ in formula (11.10):

$$\widetilde{A}\Psi(z) = \left(\frac{1}{2\pi\hbar}\right)^n \int_{\mathbb{R}^{2n}} a_\sigma(z_0)\widetilde{T}(z_0)\Psi(z)dz_0. \tag{11.11}$$

We note that the operators $\widetilde{T}(z_0)$ can be extended to operators $\widetilde{T}(z_0) :$ $S'(\mathbb{R}^{2n}) \longrightarrow S'(\mathbb{R}^{2n})$ whose restrictions to $L^2(\mathbb{R}^{2n})$ are unitary, and that satisfy the same commutation relations

$$\widetilde{T}(z_0)\widetilde{T}(z_1) = e^{\frac{i}{\hbar}\sigma(z_0,z_1)}\widetilde{T}(z_1)\widetilde{T}(z_0) \tag{11.12}$$

as the Heisenberg–Weyl operators.

Bopp operators are Weyl operators in their own right, but with symbol defined on twice the phase space, that is on $\mathbb{R}^{2n} \times \mathbb{R}^{2n} \equiv \mathbb{R}^{4n}$.

**Proposition 143.** *Let* $\widetilde{A} = \mathrm{Op}_\mathrm{B}(a)$. *We have* $\widetilde{A} = \mathrm{Op}_\mathrm{W}(\widetilde{a})$ *where the symbol* $\widetilde{a} \in S'(\mathbb{R}^{4n})$ *is defined by*

$$\widetilde{a}(z,\zeta) = a\left(z - \frac{1}{2}J\zeta\right). \tag{11.13}$$

We will not prove this result here, and we refer to de Gosson [39, Chapter 18, §18.1.2]. The idea of the proof is to determine the distributional kernel of $\widetilde{A}$ and to thereafter to calculate the corresponding Weyl symbol $\widetilde{a}$.

Observe that if we set $\zeta = (\zeta_x, \zeta_p)$ formula (11.13) becomes

$$\widetilde{a}(z,\zeta) = a\left(x - \frac{1}{2}\zeta_p, p + \frac{1}{2}\zeta_x\right). \tag{11.14}$$

This can be used to give a heuristic justification for the notation

$$\widetilde{A} = a\left(x + \frac{1}{2}i\hbar\partial_p, p - \frac{1}{2}i\hbar\partial_x\right)$$

in terms of Bopp shifts. This formula is very much used in the physical literature as a definition; one should however be aware of the fact that it is only a formal expression. For instance, when $a$ is a Hamiltonian function $(n = 1)$

$$H(x, p) = \frac{1}{2m}p^2 + V(x),$$

physicists write the corresponding Bopp operator as

$$\widetilde{H} = \frac{1}{2m}\left(x + \frac{1}{2}i\hbar\frac{\partial}{\partial p}\right)^2 + V\left(p - \frac{1}{2}i\hbar\frac{\partial}{\partial x}\right).$$

While the first term in the right-hand side has a precise meaning as a differential operator, the second term arising from the potential $V$ is a priori meaningless. This difficulty is often circumvented in physics by expanding formally the potential in a Taylor series, and thereafter replacing the variable $x$ with the Bopp shift $p - \frac{1}{2}i\hbar\partial/\partial x$. Of course such a procedure is only acceptable when the function $V$ is the restriction to $\mathbb{R}$ of an analytic function.

**Remark 144.** An immediate consequence of formula (11.13) is the following: If we change the symbol $a$ into its complex conjugate, then both $\widehat{A}$ and $\widetilde{A}$ become the adjoints $\widehat{A}^*$ and $\widetilde{A}^*$. In particular $\widetilde{A}$ is self-adjoint if and only if $a$ (and hence $\widetilde{a}$) is real.

### 11.2.3. The intertwining property

The wavepacket transforms $U_\phi$ introduced in Chapter 6 play the role of "intertwiners" between the usual Weyl operators and the Bopp operators. Recall that by definition

$$U_\phi\psi(z) = (2\pi\hbar)^{n/2}W(\psi, \phi)(z).$$

We begin by noting the following (almost obvious) result.

**Lemma 145.** *Let $\widetilde{T}(z_0)$ be the unitary operator on $L^2(\mathbb{R}^{2n})$ defined by* (11.9). *We have*

$$\widetilde{T}(z_0)U_\phi\psi = U_\phi(\widehat{T}(z_0)\psi). \tag{11.15}$$

**Proof.** In Chapter 2, formula (2.32), we showed that

$$W(\widehat{T}(z_0)\psi, \phi)(z) = e^{-\frac{i}{\hbar}\sigma(z,z_0)}W(\psi, \phi)\left(z - \frac{1}{2}z_0\right);$$

equivalently

$$U_\phi(\widehat{T}(z_0)\psi) = e^{-\frac{i}{\hbar}\sigma(z,z_0)}U_\phi\psi\left(z - \frac{1}{2}z_0\right)$$

which is (11.15). $\qquad\square$

Applying this lemma to Bopp operators yields the following.

**Proposition 146.** *Let* $\widehat{A} = \mathrm{Op_W}(a)$ *and* $\widetilde{A} = \mathrm{Op_B}(a)$. *We have*

$$\widetilde{A}U_\phi = U_\phi\widehat{A} \quad \text{and} \quad U_\phi^*\widetilde{A} = \widehat{A}U_\phi^*. \tag{11.16}$$

**Proof.** Applying $U_\phi$ to both sides of the formula

$$\widehat{A}\psi = \left(\frac{1}{2\pi\hbar}\right)^n \int_{\mathbb{R}^{2n}} a_\sigma(z_0)\widehat{T}(z_0)\psi dz_0,$$

we get, using formula (11.15),

$$U_\phi\widehat{A}\psi = \left(\frac{1}{2\pi\hbar}\right)^n \int_{\mathbb{R}^{2n}} a_\sigma(z_0)U_\phi[\widehat{T}(z_0)\psi]dz_0$$

$$= \left(\frac{1}{2\pi\hbar}\right)^n \int_{\mathbb{R}^{2n}} a_\sigma(z_0)[\widetilde{T}(z_0)U_\phi\psi]dz_0$$

$$= \widetilde{A}U_\phi\psi$$

hence the first equality (11.16). Replacing $\widetilde{A}$ with $\widetilde{A}^*$ has the effect of replacing $\widehat{A}$ with $\widehat{A}^*$ (see Remark 144) so we have $\widetilde{A}^*U_\phi = U_\phi\widehat{A}^*$. Taking the adjoint of both sides of this equality, we get $U_\phi^*\widetilde{A} = U_\phi\widehat{A}^*$ which is the second equality (11.16). $\qquad\square$

Using Proposition 146 one can relate the spectral properties of the Bopp operator $\widetilde{A} = \mathrm{Op_B}(a)$ to those of the Weyl operator $\widehat{A} = \mathrm{Op_W}(a)$: see [37, 39, 47] for detailed results, including applications to the Landau problem.

**Main Formulas in Chapter 11**

| | |
|---|---|
| Moyal product $a \star_\hbar b$ | $\mathrm{Op_W}(a \star_\hbar b) = \mathrm{Op_W}(a)\mathrm{Op_W}(b)$ |
| Explicit expression | $\left(\frac{1}{\pi\hbar}\right)^{2n} \int_{\mathbb{R}^{4n}} e^{-\frac{2i}{\hbar}\sigma(u-z,v-z)}a(u)b(v)dudv$ |
| twisted convolution $a\#b$ | $a\#b = F_\sigma(F_\sigma a \star_\hbar F_\sigma b)$ |
| explicit expression | $\left(\frac{1}{2\pi\hbar}\right)^n \int_{\mathbb{R}^{2n}} e^{\frac{i}{2\hbar}\sigma(z,z')}a_\sigma(z-z')b_\sigma(z')dz'$ |
| Bopp shifts | $x_j\star_\hbar = x_j + \frac{i\hbar}{2}\frac{\partial}{\partial p_j}, \quad p\star_\hbar = p_j - \frac{i\hbar}{2}\frac{\partial}{\partial x_j}$ |
| operator $\widetilde{T}(z_0)$ | $\widetilde{T}(z_0)\Psi(z) = e^{-\frac{i}{\hbar}\sigma(z,z_0)}\Psi(z - \frac{1}{2}z_0)$ |
| Bopp operator | $\widetilde{A} = \left(\frac{1}{2\pi\hbar}\right)^n \int_{\mathbb{R}^{2n}} a_\sigma(z_0)\widetilde{T}(z_0)dz_0$ |
| WPT | $U_\phi\psi = (2\pi\hbar)^{n/2}W(\psi,\phi)$ |
| Intertwining relations | $\widetilde{A}U_\phi = U_\phi\widehat{A} \quad \text{and} \quad U_\phi^*\widetilde{A} = \widehat{A}U_\phi^*$ |

Chapter 12

# Probabilistic Interpretation of the Wigner Transform

**Summary 147.** The Wigner transform plays a role perfectly similar to that of a probability distribution in quantum mechanics. Using the Wigner transform, we can define the covariance matrix and prove a strong version of the uncertainty principle, due to Robertson and Schrödinger. The notion of weak values of an observable (= symbol) is also introduced.

This chapter in a sense the key to the quantum mechanical probabilistic interpretation of the Wigner transform, as meant by Wigner himself in his foundational paper [81].

## 12.1. Introduction

We begin by shortly collecting some properties of the Wigner transform we have proved in the previous chapters. We thereafter introduce the fundamental notion of expectation value.

### 12.1.1. Back to Wigner

Recall that the Wigner transform of a function $\psi \in L^2(\mathbb{R}^n)$ is continuous on $\mathbb{R}^{2n}$ and satisfies

$$\int_{\mathbb{R}^{2n}} W\psi(z)dz = ||\psi||_{L^2}$$

(formula (2.23) in Chapter 2). It follows that when $\psi$ is normalized $(||\psi||_{L^2} = 1)$ then

$$\int_{\mathbb{R}^{2n}} W\psi(z)dz = 1$$

which is one of the essential properties which should be verified by a probability distribution. However, we have seen that $W\psi$ usually takes negative values; for this reason it is common in the literature to call $W\psi$ a quasi-probability distribution (or, simply, a quasi-distribution). In the rest of this chapter we will show that this probabilistic interpretation of the Wigner transform can be pushed quite far.

We also recall that we have proven the Wigner transform satisfies the "right" marginal properties, in the sense that if $\psi$ is both integrable and square integrable then

$$\int_{\mathbb{R}^n} W\psi(x,p)dp = |\psi(x)|^2,$$

$$\int_{\mathbb{R}^n} W\psi(x,p)dx = |F\psi(p)|^2$$

(formulas (2.26) and (2.27) in Chapter 2). In fact, the marginals $|\psi(x)|^2$ and $|F\psi(p)|^2$ are (when $||\psi||_{L^2} = 1$) the probability distributions for finding a quantum system represented by the state $\psi$ in a certain region of configuration space $\mathbb{R}^n_x$ while having its momentum in some region of momentum space $\mathbb{R}^n_p$ (see any textbook on elementary quantum mechanics). The essence of the quantum uncertainty principle is that both regions cannot be simultaneously arbitrarily small.

### 12.1.2. Averaging observables and symbols

In the language of quantum mechanics, an observable is a real function defined on phase space, and which corresponds to some physical quantity (position, momentum, energy, ...). We will be viewing here observables in a much more general sense: for us an observable is a function or distribution defined on phase space $\mathbb{R}^{2n}$; we do not necessarily require them to be real functions (hence the associated operators are not required to be self-adjoint). Mathematically speaking, observables are viewed as pseudodifferential symbols.

**Definition 148.** Let variable $a$ be a real observable, and $\widehat{A} = \mathrm{Op_W}(a)$ the associated Weyl operator. Let $\psi \in L^2(\mathbb{R}^n)$ $(||\psi||_{L^2} \neq 0)$ and assume that

$\widehat{A}\psi \in L^2(\mathbb{R}^n)$. The expectation value (one also says average, or mean value) of $\widehat{A}$ in the state $\psi$ is the real number

$$\langle\widehat{A}\rangle_\psi = \frac{(\widehat{A}\psi|\psi)_{L^2}}{(\psi|\psi)_{L^2}}.$$

When no confusion is likely to arise, we will write simply $\langle\widehat{A}\rangle$ instead of $\langle\widehat{A}\rangle_\psi$.

The symbol $a$ being real, the Weyl operator is essentially self-adjoint, hence $(\widehat{A}\psi|\psi)_{L^2} = (\psi|\widehat{A}\psi)_{L^2}$; since $(\widehat{A}\psi|\psi)_{L^2} = \overline{(\psi|\widehat{A}\psi)_{L^2}}$ it follows that $(\widehat{A}\psi|\psi)_{L^2}$, and hence $\langle\widehat{A}\rangle_\psi$, is real.

**Remark 149.** In the quantum mechanical literature, one often finds the definition $\langle\widehat{A}\rangle_\psi = \langle\psi|\widehat{A}|\psi\rangle$ where $\psi$ is normalized. This definition is perfectly equivalent to the one above: we have, by definition, $\langle\psi|\widehat{A}|\psi\rangle = \langle\psi|\widehat{A}\psi\rangle = (\widehat{A}\psi|\psi)_{L^2}$.

There is a fundamental relationship between the average $\langle\widehat{A}\rangle_\psi$ and the Wigner transform $W\psi$. It is a consequence of Proposition 49 in Chapter 4, where we proved the following connection between Weyl operators, their symbols, and the cross-Wigner transform:

$$(\widehat{A}\psi|\phi)_{L^2} = \int_{\mathbb{R}^{2n}} a(z)W(\psi,\phi)(z)dz. \tag{12.1}$$

When $\psi, \phi \in \mathcal{S}(\mathbb{R}^n)$ then $W(\psi,\phi) \in \mathcal{S}(\mathbb{R}^{2n})$ so we can write this equality in the form

$$(\widehat{A}\psi|\phi)_{L^2} = \langle a, W(\psi,\phi)\rangle. \tag{12.2}$$

This equality often holds under more general conditions on the functions $\psi$ and $\phi$; for instance if $\widehat{A}$ is bounded on $L^2(\mathbb{R}^n)$ it is true for all $\psi, \phi \in L^2(\mathbb{R}^n)$.

**Proposition 150.** *The expectation value of the Weyl operator* $\widehat{A} = \text{Op}_W(a)$ *in the state* $\psi \neq 0$ *is given by*

$$\langle\widehat{A}\rangle_\psi = \frac{1}{(\psi|\psi)_{L^2}} \int_{\mathbb{R}^{2n}} a(z)W\psi(z)dz. \tag{12.3}$$

**Proof.** It immediately follows from Definition 148 using formula (12.1) with $\psi = \phi$. □

**Remark 151.** Formula (12.1) will also play an essential role when we deal with the more general notion of weak value.

## 12.2. The Strong Uncertainty Principle

The strong uncertainty principle, also called the Robertson–Schrödinger principle, is a generalization of the textbook Heisenberg inequality $\Delta p \Delta x \geq \frac{1}{2}\hbar$.

### 12.2.1. Variances and covariances

In what follows $\psi \in L^1(\mathbb{R}^n) \cap L^2(\mathbb{R}^n)$ is normalized: $||\psi||_{L^2} = 1$; setting $\rho = W\psi$ we have

$$\int_{\mathbb{R}^{2n}} \rho(z)dz = 1 \qquad (12.4)$$

and the marginal identities (2.26) and (2.27) hold:

$$\int_{\mathbb{R}^n} \rho(x,p)dp = |\psi(x)|^2, \quad \int_{\mathbb{R}^n} \rho(x,p)dx = |F\psi(p)|^2.$$

We will in addition assume that $\rho$ decreases sufficiently fast at infinity, so that

$$\int_{\mathbb{R}^{2n}} (1 + |z|^2)|\rho(z)|dz < \infty. \qquad (12.5)$$

Let us introduce the following notation: we set $z_\alpha = x_\alpha$ if $1 \leq \alpha \leq n$ and $z_\alpha = p_{\alpha-n}$ if $n + 1 \leq \alpha \leq 2n$.

**Definition 152.** The covariances and variances of the variables $z = (x,p)$ associated with $\rho$ are the numbers

$$\Delta(z_\alpha, z_\beta) = \int_{\mathbb{R}^{2n}} (z_\alpha - \langle z_\alpha \rangle)(z_\beta - \langle z_\beta \rangle)\rho(z)dz \qquad (12.6)$$

and

$$(\Delta z_\alpha)^2 = \Delta(z_\alpha, z_\alpha) = \int_{\mathbb{R}^{2n}} (z_\alpha - \langle z_\alpha \rangle)^2 \rho(z)dz. \qquad (12.7)$$

In the formulas (12.6) and (12.7) above $\langle z_\alpha \rangle$ is the averages with respect to $\rho$ of $z_\alpha$; more generally one defines for an integer $k \geq 0$ the moments

$$\langle z_\alpha^k \rangle = \int_{\mathbb{R}^{2n}} z_\alpha^k \rho(z)dz. \qquad (12.8)$$

Notice that our condition (12.5) guarantees the existence of both $\langle z_\alpha \rangle$ and $\langle z_\alpha^2 \rangle$ as follows from the trivial inequalities

$$\left| \int_{\mathbb{R}^{2n}} z_\alpha \rho(z) dz \right| \leq \int_{\mathbb{R}^{2n}} (1 + |z|^2) |\rho(z)| dz < \infty,$$

$$\left| \int_{\mathbb{R}^{2n}} z_\alpha z_\beta \rho(z) dz \right| \leq \int_{\mathbb{R}^{2n}} (1 + |z|^2) |\rho(z)| dz < \infty.$$

It follows that the quantities (12.6) and (12.7) are well defined in view of condition (12.5).

Since the integral of $\rho$ is equal to one, formulae (12.6) and (12.7) can be rewritten as

$$\Delta(z_\alpha, z_\beta) = \langle z_\alpha z_\beta \rangle - \langle z_\alpha \rangle \langle z_\beta \rangle, \tag{12.9}$$

$$(\Delta z_\alpha)^2 = \Delta(z_\alpha, z_\alpha) = \langle z_\alpha^2 \rangle - \langle z_\alpha \rangle^2. \tag{12.10}$$

**Definition 153.** We will call the symmetric $2n \times 2n$ matrix

$$\Sigma = [\Delta(z_\alpha, z_\beta)]_{1 \leq \alpha, \beta \leq 2n}$$

the covariance matrix associated with $\rho$. When $\det \Sigma \neq 0$ the inverse $\Sigma^{-1}$ is called the precision (or information) matrix.

For instance, when $n = 1$, the covariance matrix is simply

$$\Sigma = \begin{pmatrix} \Delta x^2 & \Delta(x, p) \\ \Delta(p, x) & \Delta p^2 \end{pmatrix},$$

where the quantities $\Delta x^2$ and $\Delta(x, p)$ are defined by

$$\Delta x^2 = \langle x^2 \rangle - \langle x \rangle^2, \quad \Delta p^2 = \langle p^2 \rangle - \langle p \rangle^2,$$

$$\Delta(x, p) = \langle xp \rangle - \langle x \rangle \langle p \rangle.$$

### 12.2.2. The uncertainty principle

The textbook form

$$\Delta p_j \Delta x_j \geq \frac{1}{2} \hbar \tag{12.11}$$

of the quantum uncertainty principle is not very tractable because it does not transform well under symplectic transformations. It is preferable for all practical (and theoretical) purposes to use the stronger formulation

$$(\Delta x_j)^2 (\Delta p_j)^2 \geq \Delta(x_j, p_j)^2 + \frac{1}{4}\hbar^2; \qquad (12.12)$$

these inequalities are called the Robertson–Schrödinger inequalities in the literature; they imply the Heisenberg inequalities (12.11) when the covariances $\Delta(x_j, p_j)$ are neglected. These inequalities can be directly derived from the probabilistic interpretation of the function $\psi$. We will not do this here, but rather show that they are a consequence of a certain "quantum condition" on the covariance matrix. (A generalized Robertson–Schrödinger uncertainty principle, valid for arbitrary pairs of non-commuting operators will be proven by a direct calculation in Chapter 13.)

### 12.2.3. The quantum covariance matrix

We will be dealing with the complex matrix $\Sigma + \frac{i\hbar}{2}J$ where $J$ is the standard symplectic matrix. We note that this matrix is Hermitian:

$$\left(\Sigma + \frac{i\hbar}{2}J\right)^* = \Sigma + \frac{i\hbar}{2}J$$

since $\Sigma$ is real symmetric and $(iJ)^* = (-i)(-J) = iJ$.

The following definition will be motivated later on (Chapter 14); it characterizes quantum mechanical covariance matrices.

**Definition 154.** A $2n \times 2n$ covariance matrix $\Sigma$ will be called a "quantum covariance matrix" if it satisfies the condition

$$\Sigma + \frac{i\hbar}{2}J \geq 0 \qquad (12.13)$$

(which means that the Hermitian matrix $\Sigma + \frac{i\hbar}{2}$ has no negative eigenvalues).

The following result is essential, as it restates the uncertainty principle of quantum mechanics in a very concise form.

**Proposition 155.** *Let $\Sigma$ be a quantum covariance matrix. Then we have the following conditions.*

(i) *The covariance matrix $\Sigma$ is positive definite $\Sigma > 0$.*
(ii) *The Robertson–Schrödinger inequalities*

$$(\Delta x_j)^2 (\Delta p_j)^2 \geq \Delta(x_j, p_j)^2 + \frac{1}{4}\hbar^2 \qquad (12.14)$$

*hold for $1 \leq j \leq n$.*

**Proof.** (i) Let us show that $\Sigma$ is positive definite. Suppose first $\Sigma$ has an eigenvalue $\lambda < 0$, and let $z_\lambda$ be a real eigenvector corresponding to $\lambda$. Since $Jz_\lambda \cdot z_\lambda = 0$ (because $z_\lambda$ and $Jz_\lambda$ are orthogonal) we have

$$\left( \Sigma + \frac{i\hbar}{2} Jz_\lambda \right) \cdot z_\lambda = \Sigma z_\lambda \cdot z_\lambda = \lambda |z_\lambda|^2 < 0$$

which contradicts the assumption $\Sigma + \frac{i\hbar}{2} J \geq 0$. We next show that the case $\lambda = 0$ is excluded as well. Suppose indeed that 0 is an eigenvalue, and let $z_0$ be a corresponding real eigenvector. For $\varepsilon > 0$ set $z(\varepsilon) = (I + i\varepsilon J)z_0$. Using the relations $\Sigma z_0 = 0$ and $Jz_0 \cdot z_0 = \sigma(z_0, z_0) = 0$ we get, after a few calculations,

$$\left( \Sigma + \frac{i\hbar}{2} Jz_\lambda \right) z(\varepsilon) \cdot z(\varepsilon) = \varepsilon \frac{1}{2} \hbar |z_0|^2 + \varepsilon^2 \Sigma Jz_0 \cdot Jz_0.$$

Choose now $\varepsilon < 0$; then $\varepsilon \frac{1}{2} \hbar |z_0|^2 < 0$ and if $|\varepsilon|$ is small enough we have

$$\left( \Sigma + \frac{i\hbar}{2} Jz_\lambda \right) z(\varepsilon) \cdot z(\varepsilon) < 0$$

contradicting the fact that $\Sigma + \frac{i\hbar}{2} J \geq 0$. (ii) The non-negativity of the Hermitian matrix $\Sigma + \frac{i\hbar}{2} J$ can be expressed in terms of the submatrices

$$\Sigma_j = \begin{pmatrix} (\Delta x_j)^2 & \Delta(x_j, p_j) + \dfrac{i\hbar}{2} \\ \Delta(p_j, x_j) - \dfrac{i\hbar}{2} & (\Delta p_j)^2 \end{pmatrix}$$

which are non-negative provided that $\Sigma + \frac{i\hbar}{2} J$ is non-negative (Sylvester's criterion). Since

$$\mathrm{Tr}(\Sigma_j) = (\Delta x_j)^2 + (\Delta p_j)^2 \geq 0,$$

we have $\Sigma_j \geq 0$ if and only if

$$\det \Sigma_j = (\Delta x_j)^2 (\Delta p_j)^2 - \Delta(x_j, p_j)^2 - \frac{1}{4} \hbar^2 \geq 0,$$

which is equivalent to the inequality (12.14). $\qquad\square$

At this point it is appropriate to notice that (except in the case $n = 1$) the condition $\Sigma + \frac{i\hbar}{2} J \geq 0$ is *not equivalent* to the uncertainty inequalities (12.14); it is in fact a *stronger* condition: the fact that the Robertson–Schrödinger inequalities are satisfied by the entries of a covariance matrix $\Sigma$ does not imply that $\Sigma$ is a *quantum* covariance matrix, except in the case

$n = 1$. It is actually easy to see why we have equivalence when $n = 1$: the covariance matrix is just

$$\Sigma = \begin{pmatrix} \Delta x^2 & \Delta(x,p) \\ \Delta(p,x) & \Delta p^2 \end{pmatrix}$$

and since

$$\text{Tr}\left(\Sigma + \frac{i\hbar}{2}J\right) = \Delta x^2 + \Delta p^2 \geq 0$$

the condition $\Sigma + \frac{i\hbar}{2}J \geq 0$ is equivalent to $\det(\Sigma + \frac{i\hbar}{2}J) \geq 0$, that is to

$$\Delta x^2 \Delta p^2 - \left(\Delta(x,p)^2 + \frac{1}{4}\hbar^2\right) \geq 0$$

which is precisely (12.14) in the case $n = 1$. That this equivalence between $\Sigma + \frac{i\hbar}{2}J \geq 0$ and (12.14) is not true in higher dimensions is easily seen on the following counterexample. Take $n = 2$ and $\frac{1}{2}\hbar = 1$ and define a covariance matrix by

$$\Sigma = \begin{pmatrix} 1 & -1 & 0 & 0 \\ -1 & 1 & 0 & 0 \\ 0 & 0 & 1 & 0 \\ 0 & 0 & 0 & 1 \end{pmatrix}. \tag{12.15}$$

We thus have $(\Delta x_1)^2 = (\Delta x_2)^2 = 1$ and $(\Delta p_1)^2 = (\Delta p_2)^2 = 1$, and also $\Delta(x_1, p_1) = \Delta(x_2, p_2) = 0$ so that the inequalities (12.14) are trivially satisfied (they are in fact equalities). The matrix $\Sigma + iJ$ is nevertheless indefinite, hence $\Sigma$ cannot be a quantum covariance matrix.

## 12.3.  The Notion of Weak Value

The notion of weak value is relatively new in quantum mechanics. The notion can be expressed in terms of the cross-Wigner transform, which makes apparent its interpretation. The notion of weak value of a quantum observable (= operator) has been introduced in relation with "time-symmetric quantum mechanics", an alternative form of quantum mechanics exhibiting fascinating and unconventional features whose potentialities have certainly not yet been fully exploited; see for instance [2, 3].

### 12.3.1.  Definition of weak values

Following the time-symmetric approach to quantum mechanics, a quantum system is described by *two* wavefunctions $\psi$ and $\phi$ where $\phi$ evolves back from the future, while $\psi$ evolves from the past. At an instant $t$ both states interfere, and one can thus calculate the expectation value at this time using the superposition $\psi + \phi$:

$$\langle \widehat{A} \rangle_{\psi+\phi} = \frac{(\widehat{A}(\psi + \phi)|\psi + \phi)_{L^2}}{||\psi + \phi||_{L^2}}. \tag{12.16}$$

Equivalently, assuming the states $\psi$ and $\phi$ normalized to unity,

$$\langle \widehat{A} \rangle_{\psi+\phi} = \frac{1}{||\psi + \phi||_{L^2}}(\langle \widehat{A} \rangle_\phi + \langle \widehat{A} \rangle_\psi + 2\,\mathrm{Re}(\widehat{A}\psi|\phi)_{L^2}).$$

**Definition 156.** Assume $(\psi|\phi)_{L^2} \neq 0$. The complex number

$$\langle \widehat{A} \rangle_{\psi,\phi} = \frac{(\widehat{A}\psi|\phi)_{L^2}}{(\psi|\phi)_{L^2}} \tag{12.17}$$

is called the *weak value* of $\widehat{A}$ in the state $\psi, \phi$.

When $\psi = \phi$ we recover the expectation value of the operator $\widehat{A}$ as defined above:

$$\langle \widehat{A} \rangle_{\psi,\phi} = \frac{(\widehat{A}\psi|\psi)_{L^2}}{(\psi|\psi)_{L^2}} = \langle \widehat{A} \rangle_\psi.$$

We can express the weak value $\langle \widehat{A} \rangle_{\psi,\phi}$ in terms of the Weyl symbol of $\widehat{A}$ and the cross-Wigner transform of the state $\psi, \phi$.

**Proposition 157.** *Let $\widehat{A} = \mathrm{Op_W}(a)$ and $\psi, \phi \in L^2(\mathbb{R}^n)$ such that $(\psi|\phi)_{L^2} \neq 0$. The weak value $\langle \widehat{A} \rangle_{\psi,\phi}$ is given by*

$$\langle \widehat{A} \rangle_{\psi,\phi} = \frac{1}{(\psi|\phi)_{L^2}} \int_{\mathbb{R}^{2n}} a(z)W(\psi,\phi)(z)dz. \tag{12.18}$$

**Proof.** It immediately follows from the relation

$$(\widehat{A}\psi|\phi)_{L^2} = \int_{\mathbb{R}^{2n}} a(z)W(\psi,\phi)(z)dz$$

(formula (4.17) in Chapter 4). □

We can thus take formula (12.18) as an alternative definition of a weak value.

We note that Hiley [50] gives a thorough analysis of the relationship between weak values and the Moyal product from the point of view of a leading quantum physicist.

### 12.3.2. A complex phase space distribution

Assuming again that $(\psi|\phi)_{L^2} \neq 0$ let us set

$$\rho_{\psi,\phi}(z) = \frac{W(\psi,\phi)(z)}{(\psi|\phi)_{L^2}} \tag{12.19}$$

using the marginal conditions (2.26) and (2.27) in Chapter 2 satisfied by the cross-Wigner transform we get

$$\int_{\mathbb{R}^n} \rho_{\psi,\phi}(x,p)dp = \frac{\psi(x)\overline{\phi(x)}}{(\psi|\phi)_{L^2}}, \tag{12.20}$$

$$\int_{\mathbb{R}^n} \rho_{\psi,\phi}(x,p)dx = \frac{\overline{F\phi(p)}F\psi(p)}{(\psi|\phi)_{L^2}}, \tag{12.21}$$

hence the function $\rho_{\psi,\phi}$ is a complex probability quasi-distribution:

$$\int_{\mathbb{R}^{2n}} \rho_{\psi,\phi}(x,p)dpdx = 1. \tag{12.22}$$

The weak value is given in terms of that quasi-distribution by the formula

$$\langle\widehat{A}\rangle_{\psi,\phi} = \int_{\mathbb{R}^{2n}} a(z)\rho_{\psi,\phi}(z)dz. \tag{12.23}$$

**Remark 158.** In quantum mechanics, the meaning of these relations is that the readings of the pointer of the measuring device will cluster around the value

$$\mathrm{Re}\langle\widehat{A}\rangle_{\psi,\phi} = \int_{\mathbb{R}^{2n}} \mathrm{Re}(a(z)\rho_{\psi,\phi}(z))dz \tag{12.24}$$

while the quantity

$$\mathrm{Im}\langle\widehat{A}\rangle_{\psi,\phi} = \int_{\mathbb{R}^{2n}} \mathrm{Im}(a(z)\rho_{\psi,\phi}(z))dz \tag{12.25}$$

measures the shift in the variable conjugate to the pointer variable.

### 12.3.3. Reconstruction using weak values

One of the main problems in the time-symmetric formulation of quantum mechanics is to reconstruct the state $\psi$ or $\phi$ knowing the weak values of a certain observable. It turns out that this can be easily done using the results from Chapter 6. These can be reformulated as follows: suppose that the complex probability density $\rho_{\psi,\phi}$ has been determined experimentally, and that the state $\phi$ is known. Choosing $\gamma \in L^2(\mathbb{R}^n)$ such that $(\gamma|\phi)_{L^2} \neq 0$ we can then reconstruct the state $\psi$ by the formula

$$\psi(x) = 2^n \frac{(\phi|\psi)_{L^2}}{(\gamma|\phi)_{L^2}} \int_{\mathbb{R}^n} \rho_{\psi,\phi}(z_0)\widehat{R}(z_0)\gamma(x)dz_0 \qquad (12.26)$$

which is just a restatement of formula (6.10) in Proposition 69 (Chapter 6) using the definition (12.19) of $\rho_{\psi,\phi}$. For a discussion of this formula from a physical point of view, see [43].

Chapter 13

# Mixed Quantum States
# and the Density Operator

**Summary 159.** Mixed quantum states are described by Weyl operators with symbol a convex sum of Wigner transforms. These operators are called density operators. They can be identified with positive semidefinite self-adjoint trace class operators with trace one.

A pure quantum state corresponds to the datum of a single function $\psi \in L^2(\mathbb{R}^n)$ which is supposed to contain all the pertinent information to the state; in physics such a state is often denoted by "ket" $|\psi\rangle$. In practice, one is led to consider probabilistic mixtures of pure states; these are called "mixed states" in the literature. Suppose for instance that we have the choice between a number of states, described by (normalized) square integrable functions $\psi_1, \psi_2, \ldots$, each $\psi_j$ having a probability $\alpha_j$ to be the "true" state of the system under consideration (thus $\sum_j \alpha_j = 1$). The datum of the set of all pairs $(\psi_j, \alpha_j)_j$ is a mixed (quantum) state. To each mixed state $(\psi_j, \alpha_j)_j$ corresponds a self-adjoint positive trace class operator with trace equal to one; that operator is called the *density operator* of $(\psi_j, \alpha_j)_j$. A *caveat*: a mixed state $(\psi_j, \alpha_j)_j$ is *not* identical with the *superposition* $\sum_j \alpha_j \psi_j$.

## 13.1. Trace Class Operators

We collect in this preliminary section the definition and the main properties of trace-class operators. For a complete treatment see for instance [74, §6.4]; a detailed summary is given in [39, Chapter 12].

### 13.1.1. Definition and general properties

There are several equivalent definitions of the notion of trace-class operator. The following is the most flexible, because it is also the most general; it applies to arbitrary Hilbert spaces as well.

**Definition 160.** An bounded operator $\widehat{A}$ on $L^2(\mathbb{R}^n)$ is said to be of trace class if there exist two orthonormal bases $(\psi_j)$ and $(\phi_k)_k$ of such that

$$\sum_{j,k} |(\widehat{A}\psi_j|\phi_k)_{L^2}| < \infty. \tag{13.1}$$

One shows that if (13.1) holds for one pair of orthonormal bases $(\psi_j)$ and $(\phi_k)_k$ then it holds for all pairs of orthonormal bases. The trace of $\widehat{A}$ is then defined by the absolutely convergent series

$$\text{Tr}(\widehat{A}) = \sum_j (\widehat{A}\psi_j|\psi_j)_{L^2} \tag{13.2}$$

and one proves that the right-hand side of (13.2) does not depend on the choice of the basis $(\psi_j)$.

Obviously $\widehat{A}$ is of trace class if and only if its adjoint $\widehat{A}^*$ is of trace class and we have $\text{Tr}(\widehat{A}^*) = \overline{\text{Tr}(\widehat{A})}$. In particular $\text{Tr}(\widehat{A})$ is real if $\widehat{A}$ is self-adjoint.

**Proposition 161.** *Let $\widehat{A}$ be a bounded operator on $L^2(\mathbb{R}^n)$. The following statements are equivalent:*

(i) $\widehat{A}$ *is of trace class.*
(ii) *The modulus $|\widehat{A}| = (\widehat{A}^*\widehat{A})^{1/2}$ is of trace class.*
(iii) $\widehat{A}$ *is the product of two Hilbert–Schmidt operators.*

We refer to any textbook on functional analysis for a detailed proof; see e.g. the appendix in [76]. Recall that a bounded operator on $L^2(\mathbb{R}^n)$ is called a Hilbert–Schmidt operator if for some orthonormal basis $(\psi_j)$ of $L^2(\mathbb{R}^n)$ we have

$$\sum_j (\widehat{A}\psi_j|\widehat{A}\psi_j)_{L^2} = \sum_j ||\widehat{A}\psi_j||_{L^2} < \infty$$

in which case the series $\sum_j ||\widehat{A}\psi_j||_{L^2}$ is convergent for all orthonormal bases. That every trace class operator is the product of two Hilbert–Schmidt operators is easy to see using the polar decomposition of $\widehat{A}$: writing $\widehat{A} = \widehat{U}(\widehat{A}^*\widehat{A})^{1/2}$ we have $\widehat{A} = \widehat{B}^*\widehat{C}$ with $\widehat{B}^* = \widehat{U}(\widehat{A}^*\widehat{A})^{1/4}$ and $\widehat{C} = (\widehat{A}^*\widehat{A})^{1/4}$ and one easily checks that $\widehat{B}^*$ and $\widehat{C}$ are Hilbert–Schmidt operators. In particular every trace class operator on $L^2(\mathbb{R}^n)$ is the product of two operators with $L^2$ kernels. Conversely, it is immediately follows from the definitions that if $\widehat{A}$ and $\widehat{B}$ are Hilbert–Schmidt operators then $\widehat{A}\widehat{B}$ is of trace-class.

### 13.1.2. The case of Weyl operators

We now consider the case where the bounded operator $\widehat{A}$ on $L^2(\mathbb{R}^n)$ is defined by its Weyl symbol: $\widehat{A} = \mathrm{Op_W}(a)$.

**Proposition 162.** *Let $\widehat{A} = \mathrm{Op_W}(a)$ and $\widehat{B} = \mathrm{Op_W}(b)$ be Hilbert–Schmidt operators. We then have*

$$\mathrm{Tr}(\widehat{A}\widehat{B}) = \mathrm{Tr}(\widehat{B}\widehat{A}) = \left(\frac{1}{2\pi\hbar}\right)^n \int_{\mathbb{R}^{2n}} a(z)b(z)dz. \tag{13.3}$$

**Proof.** Since $\widehat{A}\widehat{B}$ is of trace class, $\mathrm{Tr}(\widehat{A}\widehat{B})$ and $\mathrm{Tr}(\widehat{B}\widehat{A})$ are well defined and finite. Since the distributional kernels of $\widehat{A}$ and $\widehat{B}$ being square integrable, it follows from Proposition 47 of Chapter 4 that we have $a \in L^2(\mathbb{R}^{2n})$ and $b \in L^2(\mathbb{R}^{2n})$. Let $(\psi_j)_j$ be an orthonormal basis of $L^2(\mathbb{R}^n)$; then, by definition

$$\mathrm{Tr}(\widehat{A}\widehat{B}) = \sum_j (\widehat{A}\widehat{B}\psi_j|\psi_j)_{L^2} = \sum_j (\widehat{B}\psi_j|\widehat{A}^*\psi_j)_{L^2}.$$

Expanding $\widehat{B}\psi_j$ and $\widehat{A}^*\psi_j$ in the basis $(\psi_j)_j$ we have

$$\widehat{B}\psi_j = \sum_k (\widehat{B}\psi_j|\psi_k)_{L^2}\psi_k \quad \text{and} \quad \widehat{A}^*\psi_j = \sum_\ell (\psi_j|\widehat{A}\psi_\ell)_{L^2}\psi_\ell$$

and hence, by the Bessel identity,

$$(\widehat{B}\psi_j|\widehat{A}^*\psi_j)_{L^2} = \sum_k (\widehat{B}\psi_j|\psi_k)_{L^2}\overline{(\widehat{A}\psi_j|\psi_k)_{L^2}}.$$

In view of the fundamental relation between Weyl operators and the cross-Wigner transform (Proposition 49 of Chapter 4), we have

$$(\widehat{A}\psi_j|\psi_k)_{L^2} = \int_{\mathbb{R}^{2n}} a(z)W(\psi_j,\psi_k)(z)dz = (a|W(\psi_k,\psi_j))_{L^2(\mathbb{R}^{2n})},$$

$$(\widehat{B}\psi_j|\psi_k)_{L^2} = \int_{\mathbb{R}^{2n}} b(z)W(\psi_j,\psi_k)(z)dz = (b|W(\psi_k,\psi_j))_{L^2(\mathbb{R}^{2n})}$$

and hence

$$(\widehat{B}\psi_j|\widehat{A}^*\psi_j)_{L^2} = \sum_{k=1}^{\infty}(a|W(\psi_k,\psi_j))_{L^2(\mathbb{R}^{2n})}(b|W(\psi_k,\psi_j))_{L^2(\mathbb{R}^{2n})}.$$

Recall now from Proposition 75 of Chapter 6 that if $(\psi_j)_j$ is an orthonormal basis of $L^2(\mathbb{R}^n)$ then the functions $\Phi_{j,k} = (2\pi\hbar)^{n/2}W(\psi_k,\psi_j)$ form an orthonormal basis of $L^2(\mathbb{R}^{2n})$; thus

$$\text{Tr}(\widehat{A}\widehat{B}) = \sum_{j=1}^{\infty}(\widehat{B}\psi_j|\widehat{A}^*\psi_j)_{L^2(\mathbb{R}^n)}$$

$$= \left(\frac{1}{2\pi\hbar}\right)^n \sum_{1\leq j,k<\infty}(a|\Phi_{j,k})_{L^2(\mathbb{R}^{2n})}(b|\Phi_{j,k})_{L^2(\mathbb{R}^{2n})}$$

$$= \left(\frac{1}{2\pi\hbar}\right)^n (a|\overline{b})_{L^2(\mathbb{R}^{2n})},$$

where the third equality follows from the second using Bessel's identity; equivalently

$$\text{Tr}(\widehat{A}\widehat{B}) = \left(\frac{1}{2\pi\hbar}\right)^n \int_{\mathbb{R}^{2n}} a(z)b(z)dz$$

which is formula (13.3).                                                        $\square$

The following result expresses the trace in terms of the symbol provided a mild integrability condition is satisfied (see [20]).

**Proposition 163.** *Let $\widehat{A} = \text{Op}_{\text{W}}(a)$ be a trace class operator. If in addition $a \in L^1(\mathbb{R}^n)$ then*

$$\text{Tr}(\widehat{A}) = \left(\frac{1}{2\pi\hbar}\right)^n \int_{\mathbb{R}^{2n}} a(z)dz, \qquad (13.4)$$

*where $F$ is the $\hbar$-Fourier transform on $\mathbb{R}^{2n}$; equivalently*

$$\text{Tr}(\widehat{A}) = Fa(0). \qquad (13.5)$$

**Proof.** The equivalence of formulas (13.4) and (13.5) is clear. Writing $\widehat{A} = \widehat{B}\widehat{C}$ where $\widehat{B}$ and $\widehat{C}$ are Hilbert–Schmidt operators we have, using formula (13.3) in Proposition 162,

$$\text{Tr}(\widehat{A}) = \left(\frac{1}{2\pi\hbar}\right)^n \int_{\mathbb{R}^{2n}} b(z)c(z)dz.$$

Let us show that

$$Fa(0) = \left(\frac{1}{2\pi\hbar}\right)^n \int_{\mathbb{R}^{2n}} b(z)c(z)dz;$$

formula (13.4) will follow. In view of formula (11.6) in Proposition 139 (Chapter 11), the twisted symbol $a_\sigma = F_\sigma a$ of $\widehat{A}$ is given by

$$a_\sigma(z) = \left(\frac{1}{2\pi\hbar}\right)^n \int_{\mathbb{R}^{2n}} e^{\frac{i}{2\hbar}\sigma(z,z')} b_\sigma(z - z')c_\sigma(z')dz'$$

and hence

$$a_\sigma(0) = \left(\frac{1}{2\pi\hbar}\right)^n \int_{\mathbb{R}^{2n}} (b_\sigma)^\vee(z)c_\sigma(z)dz$$

$$= \left(\frac{1}{2\pi\hbar}\right)^n ((b_\sigma)^\vee|\overline{c_\sigma})_{L^2(\mathbb{R}^{2n})}$$

with $(b_\sigma)^\vee(z) = b_\sigma(-z)$. Noting that $(b_\sigma)^\vee = (b^\vee)_\sigma$ and $\overline{c_\sigma} = (\overline{c}^\vee)_\sigma$ we thus have,

$$a_\sigma(0) = \left(\frac{1}{2\pi\hbar}\right)^n ((b^\vee)_\sigma|(\overline{c}^\vee)_\sigma)_{L^2(\mathbb{R}^{2n})}$$

$$= \left(\frac{1}{2\pi\hbar}\right)^n (b^\vee|\overline{c}^\vee)_{L^2(\mathbb{R}^{2n})}$$

$$= \left(\frac{1}{2\pi\hbar}\right)^n \int_{\mathbb{R}^{2n}} b(z)c(z)dz$$

which was to be proven.                                                  □

## 13.2. The Density Operator

In this section we study the elementary property of the density operator of a mixed state $(\psi_j, \alpha_j)_j$ (density operators are also called "density matrices" in the physical literature). It is customary to define the density operator as being a self-adjoint positive semidefinite operator of trace class having trace equal to one. We prefer to give an equivalent definition which has the advantage of highlighting from the beginning the role played by the Wigner transform.

### 13.2.1. The Wigner transform of a mixed state

Let $\psi \in L^2(\mathbb{R}^n)$ be a (normalized) function. The datum of $\psi$ is equivalent (up to an unessential factor with modulus one) to the datum of its Wigner transform $W\psi$. In turn, the latter is uniquely determined by the Weyl operator: we have seen in Proposition 52 in Chapter 4 that the orthogonal projection $\widehat{\Pi}_\psi$ on the ray $\{\lambda\psi : \lambda \in \mathbb{C}\}$ is the Weyl operator with symbol

$$\pi_\psi = (2\pi\hbar)^n W\psi.$$

This remark allows us to identify a quantum state $\psi$ with its Wigner transform, or equivalently, with the orthogonal projector it determines. This motivates the following definition.

**Definition 164.** Let $(\psi_j, \alpha_j)_j$ be a mixed state: $||\psi_j||_{L^2} = 1$, $\alpha_j \geq 0$, $\sum_j \alpha_j = 1$. The density operator $\widehat{\rho}$ of this state is the Weyl operator

$$\widehat{\rho} = (2\pi\hbar)^n \sum_j \alpha_j \mathrm{Op_W}(W\psi_j). \tag{13.6}$$

The Wigner transform of the mixed state $(\psi_j, \alpha_j)_j$ is the function

$$\rho = \sum_j \alpha_j W\psi_j. \tag{13.7}$$

The density operator can equivalently be defined by the formula

$$\widehat{\rho} = \sum_{j \in \Lambda} \alpha_j \widehat{\Pi}_j \tag{13.8}$$

where $\widehat{\Pi}_j$ is the orthogonal projection on the ray $\{\lambda\psi_j : \lambda \in \mathbb{C}\}$. It is thus a convex sum of rank-one orthogonal projections. Conversely we have the following.

**Proposition 165.** *Let* $(\widehat{\Pi}_j)_j$ *be a family of orthogonal projectors with rank one and* $(\alpha_j)_j$ *a family of real nonnegative numbers such that* $\sum_j \alpha_j = 1$. *Then* $\widehat{\rho} = \sum_j \alpha_j \widehat{\Pi}_j$ *is the density operator of some mixed state* $(\psi_j, \alpha_j)_j$.

**Proof.** It suffices to prove that for every rank-one orthogonal projection $\widehat{\Pi}$ in $L^2(\mathbb{R}^n)$ there exists a normalized $\psi \in L^2(\mathbb{R}^n)$ such that

$$\widehat{\Pi} = (2\pi\hbar)^n \mathrm{Op_W}(W\psi).$$

Since $\widehat{\Pi}$ has rank one we have $\widehat{\Pi}\phi = \lambda(\phi|\psi)_{L^2}\psi$ for some $\psi \in L^2(\mathbb{R}^n)$; choosing $\psi$ normalized to unity the relation $\widehat{\Pi}^2 = \widehat{\Pi}$ implies that we must have $\lambda = 1$ hence our claim. $\square$

It immediately follows from formula (13.8) that density operators are compact operators (which is already clear since a density operator is the product of two Hilbert–Schmidt operators).

If $\widehat{\rho}$ is a density operator then $\widehat{\rho}^2$ is trace class; it is easy to show, using for instance (13.8), that

$$0 \leq \mathrm{Tr}(\widehat{\rho}^2) = \sum_j \alpha_j^2 \leq 1 \tag{13.9}$$

with equality if and only if $\rho = W\psi$ (i.e. when $\widehat{\rho}$ represents a "pure state"); the number $\mathrm{Tr}(\widehat{\rho}^2)$ is called the purity of the quantum state represented by the density operator $\widehat{\rho}$.

### 13.2.2. A characterization of density operators

Density operators

$$\widehat{\rho} = \sum_j \alpha_j \widehat{\Pi}_j = (2\pi\hbar)^n \sum_j \alpha_j \, \mathrm{Op_W}(W\psi_j) \tag{13.10}$$

can be identified with positive semidefinite and self-adjoint trace class operators with unit trace.

**Proposition 166.** *Let $\widehat{\rho}$ be a bounded operator on $L^2(\mathbb{R}^n)$.*

 (i) *Suppose $\widehat{\rho}$ is a density operator. Then it is self-adjoint, semidefinite positive, and has trace equal to one.*
 (ii) *Conversely, if $\widehat{\rho}$ is self-adjoint, semidefinite positive, and has trace equal to one, then it is a density operator.*

**Proof.** (i) Assume that $\widehat{\rho}$ is a density operator; then formula (13.10) implies that $\widehat{\rho}$ is self-adjoint since we have

$$\widehat{\rho}^* = \sum_j \alpha_j \widehat{\Pi}_j^* = \sum_j \alpha_j \widehat{\Pi}_j = \widehat{\rho}$$

(alternatively, it also follows from the fact that the Weyl symbol of $\widehat{\rho}$ is real, being a linear convex combination of Wigner transforms). That we have $\widehat{\rho} \geq 0$ is also clear since for every $\phi \in L^2(\mathbb{R}^n)$ we have $\widehat{\Pi}_j \phi = (\phi|\psi_j)\psi_j$ and hence

$$(\widehat{\rho}\phi|\phi)_{L^2} = \sum_j \alpha_j(\widehat{\Pi}_j\phi|\phi)_{L^2} = \sum_j \alpha_j|(\phi|\psi_j)|^2 \geq 0.$$

There remains to check the trace property. We have

$$(\widehat{\Pi}_j\psi_k|\psi_k) = (\psi_k|\psi_j)(\psi_j|\psi_k) = \delta_{jk}$$

and hence

$$\sum_k (\widehat{\rho}\psi_k | \psi_k) = \sum_{j,k} \alpha_j (\widehat{\Pi}_j \psi_k | \psi_k) = \sum_j \alpha_j = 1.$$

Alternatively,

$$\mathrm{Tr}(\widehat{\rho}) = \int_{\mathbb{R}^{2n}} \rho(z)dz = \sum_j \alpha_j \int_{\mathbb{R}^{2n}} W\psi_j(z)dz$$

and hence $\mathrm{Tr}(\widehat{\rho}) = 1$ since the $\psi_j$ are normalized to one, and the $\alpha_j$ sum up to one. (ii) Suppose that the bounded operator $\widehat{\rho}$ has these properties. In view of the spectral theorem we have

$$\widehat{\rho} = \sum_j \lambda_j \widehat{\Pi}_{\mathcal{H}_j}$$

where the $\lambda_j$ are the eigenvalues of $\widehat{\rho}$ and $\widehat{\Pi}_{\mathcal{H}_j}$ is the orthogonal projection on the eigenspace $\mathcal{H}_j$ corresponding to $\lambda_j$. The result easily follows (see [39, Proposition 275]).                    □

The following remark is important.

**Remark 167.** In the statement of the result above we insisted that a density operator is semidefinite positive. In fact, it can very well happen that some non-zero vector $\psi$ is orthogonal to all the $\psi_j$ and therefore $(\widehat{\rho}\psi|\psi)_{L^2} = 0$. However, if the state vectors $\psi_j$ form an (orthonormal) basis, then we cannot have $(\psi|\psi_j)_{L^2} = 0$ for $\psi \neq 0$; in this case $\widehat{\rho}$ is positive definite.

### 13.2.3. Uncertainty principle for density operators

Recall that we have introduced above (formula (13.7)) the Wigner transform of the mixed state $\{(\psi_j, \alpha_j)\}$: by definition

$$\rho = \sum_{j \in \Lambda} \alpha_j W\psi_j; \tag{13.11}$$

it is $(2\pi\hbar)^n$ times the symbol of the density operator $\widehat{\rho}$ associated with the mixed state $\{(\psi_j, \alpha_j)\}$. Also recall that in the course of the proof of Proposition 166 we showed that

$$\mathrm{Tr}(\widehat{\rho}) = \int_{\mathbb{R}^{2n}} \rho(z)dz = 1.$$

This strongly suggests that the function $\rho$ could play the role of a probability quasi-distribution, exactly as the Wigner transform does. This is actually commonplace in quantum mechanics.

Let $\widehat{A}$ and $\widehat{B}$ be Weyl operators; we assume that the expectation values

$$\langle \widehat{A} \rangle_{\widehat{\rho}} = \mathrm{Tr}(\widehat{\rho}\widehat{A}), \quad \langle \widehat{A}^2 \rangle_{\widehat{\rho}} = \mathrm{Tr}(\widehat{\rho}\widehat{A}^2), \tag{13.12}$$

$$\langle \widehat{B} \rangle_{\widehat{\rho}} = \mathrm{Tr}(\widehat{\rho}\widehat{B}), \quad \langle \widehat{B}^2 \rangle_{\widehat{\rho}} = \mathrm{Tr}(\widehat{\rho}\widehat{B}^2) \tag{13.13}$$

exist and are finite. Defining the variances and covariances by the formulas

$$(\Delta\widehat{A})^2_{\widehat{\rho}} = \langle \widehat{A}^2 \rangle_{\widehat{\rho}} - \langle \widehat{A} \rangle^2_{\widehat{\rho}}, \quad (\Delta\widehat{B})^2_{\widehat{\rho}} = \langle \widehat{B}^2 \rangle_{\widehat{\rho}} - \langle \widehat{B} \rangle^2_{\widehat{\rho}},$$

$$\Delta(\widehat{A}, \widehat{B})_{\widehat{\rho}} = \frac{1}{2}\langle \widehat{A}\widehat{B} + \widehat{B}\widehat{A} \rangle_{\widehat{\rho}} - \langle \widehat{A} \rangle_{\widehat{\rho}}\langle \widehat{B} \rangle_{\widehat{\rho}},$$

we have the following result.

**Proposition 168.** *Let $\widehat{A} = \mathrm{Op_W}(a)$ and $\widehat{B} = \mathrm{Op_W}(b)$ be two essentially self-adjoint Weyl operators on $L^2(\mathbb{R}^n)$. Assume that the expectation values (13.12) are defined. We have*

$$|\langle \widehat{A}\widehat{B} \rangle_{\widehat{\rho}}|^2 = \Delta(\widehat{A}, \widehat{B})^2_{\widehat{\rho}} - \frac{1}{4}\langle [\widehat{A}, \widehat{B}] \rangle^2_{\widehat{\rho}}, \tag{13.14}$$

*where $[\widehat{A}, \widehat{B}] = \widehat{A}\widehat{B} - \widehat{B}\widehat{A}$ and hence*

$$(\Delta\widehat{A})^2_{\widehat{\rho}}(\Delta\widehat{B})^2_{\widehat{\rho}} \geq \Delta(\widehat{A}, \widehat{B})^2_{\widehat{\rho}} - \frac{1}{4}\langle [\widehat{A}, \widehat{B}] \rangle^2_{\widehat{\rho}}. \tag{13.15}$$

*In particular if $[\widehat{A}, \widehat{B}] = i\hbar$ then*

$$(\Delta\widehat{A})^2_{\widehat{\rho}}(\Delta\widehat{B})^2_{\widehat{\rho}} \geq \Delta(\widehat{A}, \widehat{B})^2_{\widehat{\rho}} + \frac{1}{4}\hbar^2. \tag{13.16}$$

**Proof.** Replacing if necessary $\widehat{A}$ and $\widehat{B}$ by $\widehat{A} - \langle \widehat{A} \rangle_{\widehat{\rho}}$ and $\widehat{B} - \langle \widehat{B} \rangle_{\widehat{\rho}}$ we may assume that $\langle \widehat{A} \rangle_{\widehat{\rho}} = \langle \widehat{B} \rangle_{\widehat{\rho}} = 0$ so that (13.14) and (13.15) reduce to, respectively,

$$|\langle \widehat{A}\widehat{B} \rangle_{\widehat{\rho}}|^2 = \frac{1}{2}\langle \widehat{A}\widehat{B} + \widehat{B}\widehat{A} \rangle^2_{\widehat{\rho}} - \frac{1}{4}\langle [\widehat{A}, \widehat{B}] \rangle^2_{\widehat{\rho}} \tag{13.17}$$

and

$$\langle \widehat{A}^2 \rangle_{\widehat{\rho}}\langle \widehat{B}^2 \rangle_{\widehat{\rho}} \geq \frac{1}{2}\langle \widehat{A}\widehat{B} + \widehat{B}\widehat{A} \rangle^2_{\widehat{\rho}} - \frac{1}{4}\langle [\widehat{A}, \widehat{B}] \rangle^2_{\widehat{\rho}}. \tag{13.18}$$

The key of the proof now lies in writing

$$\widehat{A}\widehat{B} = \frac{1}{2}(\widehat{A}\widehat{B} + \widehat{B}\widehat{A}) + \frac{1}{2}(\widehat{A}\widehat{B} - \widehat{B}\widehat{A})$$

so that, by linearity,

$$\langle \widehat{A}\widehat{B} \rangle_{\widehat{\rho}} = \frac{1}{2}\langle \widehat{A}\widehat{B} + \widehat{B}\widehat{A} \rangle_{\widehat{\rho}} + \frac{1}{2}\langle [\widehat{A}, \widehat{B}] \rangle_{\widehat{\rho}}.$$

Now, $\langle \widehat{A}\widehat{B} + \widehat{B}\widehat{A} \rangle_{\widehat{\rho}}$ is a real number (because $\widehat{A}\widehat{B} + \widehat{B}\widehat{A}$ is self-adjoint since $\widehat{A}$ and $\widehat{B}$ are) and $\langle [\widehat{A}, \widehat{B}] \rangle_{\widehat{\rho}}$ is pure imaginary (because $[\widehat{A}, \widehat{B}]^* = -[\widehat{A}, \widehat{B}]$) hence formula (13.17). Observing that we have

$$\langle \widehat{A}\widehat{B} \rangle_{\widehat{\rho}} = \sum_j \alpha_j (\widehat{A}\widehat{B}\psi_j | \psi_j)_{L^2} = \sum_j \alpha_j (\widehat{B}\psi_j | \widehat{A}\psi_j)_{L^2} \qquad (13.19)$$

and applying the Cauchy–Schwarz inequality to each scalar product $(\widehat{B}\psi_j | \widehat{A}\psi_j)_{L^2}$ we get the inequality

$$|\langle \widehat{A}\widehat{B} \rangle_{\widehat{\rho}}|^2 \leq \sum_j \alpha_j \|\widehat{B}\psi_j\|_{L^2} \|\widehat{A}\psi_j\|_{L^2}. \qquad (13.20)$$

Since we are assuming $\langle \widehat{A} \rangle_{\widehat{\rho}}^2 = \langle \widehat{B} \rangle_{\widehat{\rho}}^2 = 0$ we have

$$\|\widehat{A}\psi_j\| = \langle \widehat{A}^2 \rangle_{\psi_j}^{1/2} = (\Delta \widehat{A})_{\psi_j}^2, \quad \|\widehat{B}\psi_j\| = \langle \widehat{B}^2 \rangle_{\psi_j}^{1/2} = (\Delta \widehat{B})_{\psi_j}^2$$

and the inequality (13.20) is thus equivalent to

$$|\langle \widehat{A}\widehat{B} \rangle_{\widehat{\rho}}| \leq \sum_j \alpha_j \langle \widehat{A}^2 \rangle_{\psi_j}^{1/2} \langle \widehat{B}^2 \rangle_{\psi_j}^{1/2}.$$

The Cauchy–Schwarz inequality immediately yields

$$|\langle \widehat{A}\widehat{B} \rangle_{\widehat{\rho}}| \leq \left( \sum_j \alpha_j \langle \widehat{A}^2 \rangle_{\psi_j}^{1/2} \right) \left( \sum_j \alpha_j \langle \widehat{B}^2 \rangle_{\psi_j}^{1/2} \right) = \langle \widehat{A}^2 \rangle_{\widehat{\rho}} \langle \widehat{B}^2 \rangle_{\widehat{\rho}}$$

hence the inequality (13.18) in view of formula (13.17).                    $\square$

In particular, when $\widehat{A} = x_j$ and $\widehat{B} = -i\hbar \partial/\partial x_j$ we obtain the Robertson–Schrödinger inequalities

$$(\Delta x_j)_{\widehat{\rho}}^2 (\Delta p_j)_{\widehat{\rho}}^2 \geq \Delta(x_j, p_j)_{\widehat{\rho}}^2 + \frac{1}{4}\hbar^2 \qquad (13.21)$$

for mixed states.

### 13.2.4. Covariance matrix

The covariance matrix is defined exactly as in Chapter 12: assuming that the moments

$$\langle z_\alpha^k \rangle_{\widehat{\rho}} = \int_{\mathbb{R}^{2n}} z_\alpha^k \rho(z) dz$$

exist for $k \leq 2$ one sets

$$\Delta(z_\alpha, z_\beta) = \langle z_\alpha z_\beta \rangle_{\widehat{\rho}} - \langle z_\alpha \rangle_{\widehat{\rho}} \langle z_\beta \rangle_{\widehat{\rho}},$$

$$(\Delta z_\alpha)^2 = \Delta(z_\alpha, z_\alpha) = \langle z_\alpha^2 \rangle_{\widehat{\rho}} - \langle z_\alpha \rangle_{\widehat{\rho}}^2.$$

The covariance matrix is then the real symmetric $2n \times 2n$ matrix

$$\Sigma_{\widehat{\rho}} = [\Delta(z_\alpha, z_\beta)]_{1 \leq \alpha, \beta \leq 2n}.$$

We will see in Chapter 14 that the covariance matrix satisfies the condition

$$\Sigma_{\widehat{\rho}} + \frac{i\hbar}{2} J \geq 0 \tag{13.22}$$

that is, $\Sigma_{\widehat{\rho}} + \frac{i\hbar}{2} J$ has no negative eigenvalues; this condition is the generalization to mixed states of the similar condition on the covariance matrix associated with a pure state $\psi \in L^2(\mathbb{R}^n)$ which we discussed in Chapter 12. It turns out that when this condition is satisfied, then the Robertson–Schrödinger inequalities (13.21) automatically hold by the same argument as in Proposition 155 of Chapter 12 using the fact that condition (13.22) implies that $\Sigma_{\widehat{\rho}}$ is positive definite. We make an important remark.

**Remark 169.** If the condition (13.22) holds for one value of $\hbar$, then it also holds for smaller values: $\Sigma_{\widehat{\rho}} + \frac{i\hbar'}{2} J \geq 0$ for every $\hbar' \leq \hbar$. In fact, setting $\hbar' = r\hbar$ with $0 < r \leq 1$ we have

$$\Sigma_{\widehat{\rho}} + \frac{i\hbar'}{2} J = (1 - r)\Sigma_{\widehat{\rho}} + r \left( \Sigma_{\widehat{\rho}} + \frac{i\hbar}{2} J \right) \geq 0$$

because $(1 - r)\Sigma \geq 0$.

The remark above is at the heart of the following natural question we will address in Chapter 14: suppose that $\widehat{\rho}$ is a density operator representing some mixed state $\{(\psi_j, \alpha_j)\}$. Suppose that we change the value of Planck's constant $h$ into another value, thus replacing $\hbar = h/2\pi$ with a new value $\hbar' = h'/2\pi$; we then obtain a new operator $\widehat{\rho}'$. Is this new operator still a density operator, and if it is how is the corresponding mixed state $\{(\psi_j', \alpha_j')\}$ related to $\{(\psi_j, \alpha_j)\}$? This question is, we will see, very far from being trivial. One might be tempted to say that the question is solved by Remark 169 above, but this is not the case: what the remark says is only that $\Sigma_{\widehat{\rho}}$ remains a quantum covariance matrix, but not of which state. Which makes things worse is that if $\widehat{\rho}$ is a (self-adjoint) trace class operator then the condition $\Sigma_{\widehat{\rho}} + \frac{i\hbar}{2} > 0$ alone is not sufficient to ensure the non-negativity of $\widehat{\rho}$: as we have pointed out in a joint paper [45] with Franz Luef the uncertainty principle does not determine the quantum state. Notice, however, that if the condition $\Sigma + \frac{i\hbar}{2} > 0$ is satisfied by some $\Sigma$ then there always exists quantum states having $\Sigma$ as a covariance matrix, namely the Gaussians

$$\psi(x) = Ce^{-\frac{1}{2}\Sigma^{-1}x\cdot x};$$

this follows from the fact that $\Sigma + \frac{i\hbar}{2} > 0$ implies that $\Sigma$ is positive definite and hence invertible.

Here is an example in dimension $n = 1$ due to Narcowich and O'Connell [72] showing that an $L^1$-function $\rho$ whose associated covariance matrix satisfies (13.22) is not necessarily a density operator.

**Example 170.** Let $\rho$ be defined by the Fourier integral

$$\int e^{i(xx'+pp)}\rho(x',p')dp'dx' = \left(1 - \frac{1}{2}(ax^2 + bp^2)\right)e^{-(a^2x^4+b^2p^4)}$$

where $a$ and $b$ are real positive numbers. The operator $\widehat{\rho}$ is self-adjoint and has trace one. One verifies that the condition $\Sigma_{\widehat{\rho}} + \frac{i\hbar}{2}J \geq 0$ holds if $ab \geq \hbar^2/4$. However, $\widehat{\rho}$ is not a density operator because it fails to be positive semidefinite; in fact in all cases

$$\int \rho(x,p)p^4dpdx = -24a^2 < 0.$$

This example and the discussion above (also see de Gosson and Luef [45]) show that positivity questions are particularly difficult to deal with the theory of mixed quantum states and density operators. We will come back to these issues in Chapter 14.

### Main Formulas in Chapter 13

| | |
|---|---|
| Trace class operator | $\sum_{j,k} \|(\widehat{A}\psi_j\|\phi_k)_{L^2}\| < \infty$ |
| Trace of an operator | $\mathrm{Tr}(\widehat{A}) = \sum_j (\widehat{A}\psi_j\|\psi_j)_{L^2}$ |
| Hilbert–Schmidt operator | $\sum_j (\widehat{A}\psi_j\|\widehat{A}\psi_j)_{L^2} < \infty$ |

| | |
|---|---|
| Trace Weyl operator | $\mathrm{Tr}(\widehat{A}) = \left(\frac{1}{2\pi\hbar}\right)^n \int_{\mathbb{R}^{2n}} a(z)dz$ |
| Trace of a product | $\mathrm{Tr}(\widehat{A}\widehat{B}) = \left(\frac{1}{2\pi\hbar}\right)^n \int_{\mathbb{R}^{2n}} a(z)b(z)dz$ |
| Density operator | $\widehat{\rho} = (2\pi\hbar)^n \sum_j \alpha_j \mathrm{Op_W}(W\psi_j)$ |
| Wigner transform of $\widehat{\rho}$ | $\rho = \sum_j \alpha_j W\psi_j$ |
| Uncertainty principle | $\Sigma_{\widehat{\rho}} + \frac{i\hbar}{2}J \geq 0$ |
| RS inequalities | $(\Delta x_j)^2_{\widehat{\rho}}(\Delta p_j)^2_{\widehat{\rho}} \geq \Delta(x_j, p_j)^2_{\widehat{\rho}} + \frac{1}{4}\hbar^2$ |

Chapter 14

# The KLM Conditions and the Narcowich–Wigner Spectrum

**Summary 171.** A density operator may lose its property of being positive semidefinite when one changes the value of Planck's constant. This is an important and difficult issue related to a quantum version of Bochner's theorem on the Fourier transform of a positive finite measure. These considerations lead to the notion of Narcowich–Wigner spectrum of a trace class operator.

In this chapter we address a difficult topic, where there still are many open problems. We have seen that a trace-class operator with trace one is a density operator (matrix) if it is in addition positive semidefinite. It turns out that it is this positivity issue which is difficult to check in practice; the only known criterion is called the "KLM conditions". In addition, positivity depends on the value of the parameter $\hbar$: a given state can be "quantum" for some values of $\hbar$, and "classical" for other. The problem of determining these "quantum values" (which constitute the "Narcowich–Wigner spectrum") for arbitrary mixed states is still open.

## 14.1. The Quantum Bochner Theorem

### 14.1.1. Bochner's theorem

Let $\rho$ be a probability density on $\mathbb{R}^m$; by definition the complex-valued function $f$ defined by

$$f(x) = \int_{\mathbb{R}^m} e^{ix \cdot y} \rho(y) dy$$

is the characteristic function of $\rho$ (or, equivalently, of the probability measure $d\mu = \rho dy$). Clearly $f(0) = 1$ and $f$ is continuous. These two conditions alone are however not sufficient to ensure that a complex-valued function $f$ is the characteristic function of a probability density. In fact, it follows from a famous theorem of Bochner [6] that such a function is a characteristic function if and only if in addition to these conditions the following positive semidefiniteness condition holds (see [63, Chapter VI] for a relatively elementary proof of Bochner's theorem).

**Proposition 172.** *The function $f$ is the Fourier transform of a positive measure on $\mathbb{R}^m$ if and only if $f$ is of positive type, that is, if for each integer $N$ the $N \times N$ complex matrix $(f(\xi_k - \xi_j)_{1 \leq j,k \leq N}$ is positive semidefinite for all $\xi_j, \xi_k \in \mathbb{R}^m$.*

The condition for $f$ to be of positive type is equivalent to the condition that for every integer $N$ we have

$$\sum_{j,k=1}^{N} \zeta_j \overline{\zeta_k} f(\xi_j - \xi_k) \geq 0$$

for all $\zeta_j, \zeta_k \in \mathbb{C}^m$ and all $\xi_j, \xi_k \in \mathbb{R}^m$.

Since we are working in phase space $\mathbb{R}^{2n}$, it will be convenient to reformulate Bochner's theorem in a slightly different way. Let us first introduce the following notation: for a function $a \in \mathcal{S}(\mathbb{R}^{2n})$ we set

$$a_\Diamond(z) = F_\Diamond a(z) = \left(\frac{1}{2\pi}\right)^n \int_{\mathbb{R}^{2n}} e^{i\sigma(z,z')} a(z') dz'.$$

Obviously, $F_\Diamond$ is related to the usual symplectic transform

$$a_\sigma(z) = F_\sigma a(z) = \left(\frac{1}{2\pi\hbar}\right)^n \int_{\mathbb{R}^{2n}} e^{-\frac{i}{\hbar}\sigma(z,z')} a(z') dz'$$

by the formula

$$a_\Diamond(z)(z) = \hbar^n a_\sigma(-\hbar z). \tag{14.1}$$

**Proposition 173.** *A real function $\rho$ on $\mathbb{R}^{2n}$ is a probability density if and only if $\rho_\Diamond(0) = (2\pi)^{-n}$ and*

$$\sum_{1 \leq j,k \leq N} \lambda_j \overline{\lambda_k} a_\Diamond(z_j - z_k) \geq 0 \tag{14.2}$$

*for all $\lambda_j, \lambda_k \in \mathbb{C}^{2n}$ and all $z_j, z_k \in \mathbb{R}^{2n}$.*

This result is of course trivial a reformulation of Bochner's theorem; note that the condition $\rho_\Diamond(0) = (2\pi)^{-n}$ is equivalent to

$$\int_{\mathbb{R}^{2n}} \rho(z)dz = 1.$$

## 14.1.2. The quantum case: the KLM conditions

The letters KLM stand for Kastler [62], and Loupias and Miracle-Sole [65, 66] whose seminal papers have contributed significantly to a better understanding of positivity issues in quantum mechanics; they are at the origin of the quantum Bochner theorem. We begin by defining the notion of $\hbar$-positivity.

**Definition 174.** Let $f$ be a complex function defined on $\mathbb{R}^{2n}$ and set

$$\Lambda_{jk}(z_j, z_k) = e^{-\frac{i\hbar}{2}\sigma(z_j, z_k)} f(z_j - z_k). \tag{14.3}$$

We say that $f$ is of $\hbar$-positive type if the matrix $\Lambda_{(N)} = (\Lambda_{jk}(z_j, z_k))_{1 \le j,k \le N}$ is positive semidefinite ($\Lambda_{(N)} \ge 0$) for all choices of $N$ and of $z_j, z_k \in \mathbb{R}^{2n}$.

The conditions $\Lambda_{(N)} \ge 0$ are equivalent to the polynomial inequalities

$$\sum_{1 \le j,k \le N} \lambda_j \overline{\lambda_k} e^{-\frac{i\hbar}{2}\sigma(z_j, z_k)} f(z_j - z_k) \ge 0$$

for all $N \in \mathbb{N}$, $\lambda_j \lambda_k \in \mathbb{C}$, and $z_j, z_k \in \mathbb{R}^{2n}$.

In what follows the function $f$ will be the reduced symplectic Fourier transform $a_\Diamond$ of a symbol $a \in \mathcal{S}'(\mathbb{R}^{2n})$. The following obvious result gives an alternative definition of the $\hbar$-positivity of $a_\Diamond$ in terms of the usual symplectic Fourier transform $a_\sigma = F_\sigma a$ when $\hbar \ne 0$.

**Lemma 175.** *The function $a_\Diamond$ is of $\hbar$-positive type ($\hbar \ne 0$) if and only if for every integer $N \ge 1$ and every $(z_1, \ldots, z_N) \in (\mathbb{R}^{2n})^N$ the $N \times N$ matrix $\Lambda' = (\Lambda'_{jk}(z_j, z_k))_{1 \le j,k \le N}$ where*

$$\Lambda'_{jk}(z_j, z_k) = e^{\frac{i}{2\hbar}\sigma(z_j, z_k)} a_\sigma(z_j - z_k) \tag{14.4}$$

*is positive semidefinite.*

**Proof.** It is immediate in view of the relation (14.1) between $a_\Diamond$ and replacing $(z_j, z_k)$ with $\hbar^{-1}(z_k, z_j)$ and noting that $\sigma(z_k, z_j) = -\sigma(z_j, z_k)$. $\qquad\square$

The right-hand side of (14.4) is reminiscent of the formula (1.10)

$$\widehat{T}(z_0 + z_1) = e^{-\frac{i}{2\hbar}\sigma(z_0,z_1)}\widehat{T}(z_0)\widehat{T}(z_1)$$

for Heisenberg–Weyl operators, which we can rewrite as

$$\widehat{T}(z_k - z_j) = e^{-\frac{i}{2\hbar}\sigma(z_j,z_k)}\widehat{T}(z_k)\widehat{T}(-z_j). \tag{14.5}$$

Let us now state the KLM conditions.

**Definition 176.** A symbol $a \in \mathcal{S}'(\mathbb{R}^{2n})$ is said to satisfy the KLM conditions if:

(i) $a_\Diamond$ is continuous and $a_\Diamond(0) = 1$;
(ii) $a_\Diamond$ is of $\hbar$-positive type.

We are next going to prove a fundamental result, namely that a function (or distribution) $\rho \in \mathcal{S}'(\mathbb{R}^{2n})$ is the Wigner transform of a density operator if and only if it satisfies the KLM conditions. We treat necessity and sufficiently separately (Propositions 177 and 180). We begin by showing that the Wigner distribution of a density matrix is a function of $\hbar$-positive type.

**Proposition 177.** *Let $\widehat{\rho}$ be a density operator on $L^2(\mathbb{R}^n)$ with Wigner distribution $\rho$. The function $\rho_\Diamond$ is of $\hbar$-positive type.*

**Proof.** By definition (13.7) we have

$$\rho = \sum_j \alpha_j W\psi_j \tag{14.6}$$

for a sequence of normalized functions $\psi_j \in L^2(\mathbb{R}^n)$, the coefficients $\alpha_j$ being $\geq 0$ and summing up to one. It is thus sufficient to show that the Wigner transform $W\psi$ of an arbitrary $\psi \in L^2(\mathbb{R}^n)$ is of $\hbar$-positive type. In view of Lemma 175 this amounts to show that for all $(z_1, \ldots, z_N) \in (\mathbb{R}^{2n})^N$ and all $(\lambda_1, \ldots, \lambda_N) \in \mathbb{C}^N$ we have

$$I_N(\psi) = \sum_{1 \leq j,k \leq N} \lambda_j \overline{\lambda_k} e^{-\frac{i}{2\hbar}\sigma(z_j,z_k)} F_\sigma W\psi(z_j - z_k) \geq 0 \tag{14.7}$$

for every complex vector $(\lambda_1, \ldots, \lambda_N) \in \mathbb{C}^N$ and every sequence $(z_1, \ldots, z_N) \in (\mathbb{R}^{2n})^N$. Since the Wigner distribution $W\psi$ and the ambiguity function

$$A\psi(z) = \left(\frac{1}{2\pi\hbar}\right)^n (\widehat{T}(z)\psi|\psi)_{L^2}$$

are obtained from each other by the symplectic Fourier transform $F_\sigma$ (formula (3.15) in Proposition 31), we have

$$I_N(\psi) = \sum_{1 \le j,k \le N} \lambda_j \overline{\lambda_k} e^{-\frac{i}{2\hbar}\sigma(z_j,z_k)} A\psi(z_j - z_k)).$$

Let us prove that

$$I_N(\psi) = \left(\frac{1}{2\pi\hbar}\right)^n \left\| \sum_{1 \le j \le N} \lambda_j \widehat{T}(z_j)\psi \right\|^2_{L^2} ; \tag{14.8}$$

the inequality (14.7) will follow. Taking into account the fact that $\widehat{T}(-z_k)^* = \widehat{T}(z_k)$ and using formula (14.5) we have, expanding the square in the right-hand side of (14.8),

$$\left\| \sum_{1 \le j \le N} \lambda_j \widehat{T}(z_j)\psi \right\|^2_{L^2} = \sum_{1 \le j,k \le N} \lambda_j \overline{\lambda_k} (\widehat{T}(z_j)\psi | \widehat{T}(z_k)\psi)_{L^2}$$

$$= \sum_{1 \le j,k \le N} \lambda_j \overline{\lambda_k} (\widehat{T}(-z_k)\widehat{T}(z_j)\psi | \psi)_{L^2}$$

$$= \sum_{1 \le j,k \le N} \lambda_j \overline{\lambda_k} e^{-\frac{i}{2\hbar}\sigma(z_j,z_k)} (\widehat{T}(z_j - z_k)\psi | \psi)_{L^2}$$

$$= (2\pi\hbar)^n \sum_{1 \le j,k \le N} \lambda_j \overline{\lambda_k} e^{-\frac{i}{2\hbar}\sigma(z_j,z_k)} A\psi(z_j - z_k)$$

proving the equality (14.8). □

**Remark 178.** The proof above shows that, more generally, if $\widehat{A} = \mathrm{Op_W}(a)$ with $a = \sum_j \alpha_j W\psi_j$ and $\alpha_j \ge 0$, $\sum_j \alpha_j < \infty$ then $\widehat{A} \ge 0$ implies that the twisted symbol $a_\sigma$ is of $\hbar$-positive type.

We are now going to prove a converse to Proposition 177. Let us first recall from Chapter 11 that if $\widehat{A} = \mathrm{Op_W}(a)$ and $\widehat{B} = \mathrm{Op_W}(b)$ then the twisted symbol $c_\sigma = F_\sigma c$ of the product $\widehat{C} = \widehat{A}\widehat{B}$ is given by the twisted convolution

$$c_\sigma(z) = \left(\frac{1}{2\pi\hbar}\right)^n \int_{\mathbb{R}^{2n}} e^{\frac{i}{2\hbar}\sigma(z,z')} a_\sigma(z - z') b_\sigma(z') dz'. \tag{14.9}$$

Also, we recall Schur's product theorem,[1] for the Hadamard product of matrices.

---

[1]See for instance R. Bapat, *Nonnegative Matrices and Applications*, Cambridge University Press, 1997.

**Lemma 179.** *Let* $A = (a_{jk})_{1 \leq j,k \leq N}$ *and* $B = (b_{jk})_{1 \leq j,k \leq N}$. *If* $A$ *and* $B$ *are positive semidefinite, then so is their Hadamard product* $A \circ B = (a_{jk}b_{jk})_{1 \leq j,k \leq N}$.

We have the following proposition.

**Proposition 180.** *Let* $\widehat{\rho} = (2\pi\hbar)^n \mathrm{Op_W}(\rho)$ *be a self-adjoint operator of trace class on* $L^2(\mathbb{R}^n)$. *If* $\rho_\diamond$ *of is of* $\hbar$-*positive type then* $\widehat{\rho} \geq 0$. *If, in addition,* $\mathrm{Tr}(\widehat{\rho}) = 1$, *then* $\widehat{\rho}$ *is thus a density operator.*

**Proof.** We have to show that $(\widehat{\rho}\psi|\psi)_{L^2} \geq 0$ for all $\psi \in L^2(\mathbb{R}^n)$; equivalently

$$\int_{\mathbb{R}^{2n}} \rho(z) W\psi(z) dz \geq 0 \qquad (14.10)$$

in view of formula (4.17) in Chapter 4. Let us set, as in Lemma 175,

$$\Lambda'_{jk} = e^{\frac{i}{2\hbar}\sigma(z_j,z_k)} \rho_\sigma(z_j - z_k), \qquad (14.11)$$

where $z_j$ and $z_k$ are arbitrary elements of $\mathbb{R}^{2n}$. To say that $\rho$ is of $\hbar$-positive type means that the matrix $\Lambda' = (\Lambda'_{jk})_{1 \leq j,k \leq N}$ is positive semidefinite; choosing $z_k = 0$ and setting $z_j = z$ this means that every matrix $(\rho_\sigma(z))_{1 \leq j,k \leq N}$ is positive semidefinite. Let us now set

$$\Gamma_{jk} = e^{\frac{i}{2\hbar}\sigma(z_j,z_k)} F_\sigma W\psi(z_j - z_k);$$

we know from Proposition 177 that the matrix $\Gamma = (\Gamma_{jk})_{1 \leq j,k \leq N}$ is positive semidefinite. Let us now write

$$\begin{aligned} M_{jk} &= F_\sigma W\psi(z_j - z_k)\rho_\sigma(z_j - z_k) \\ &= (e^{\frac{i}{2\hbar}\sigma(z_j,z_k)} F_\sigma W\psi(z_j - z_k))(e^{-\frac{i}{2\hbar}\sigma(z_j,z_k)} \rho_\sigma(z_j - z_k)). \end{aligned}$$

In view of Lemma 179 the matrix $M = (M_{jk})_{1 \leq j,k \leq N}$ is positive semidefinite since it is the Hadamard product of two positive semidefinite matrices; it follows from Bochner's theorem that the function $b$ defined by

$$b_\sigma(z) = F_\sigma W\psi(z)\rho_\sigma(-z)$$

is a probability density; in particular $b(0) \geq 0$. Integrating this equality we get, using the Plancherel formula for the symplectic Fourier transform (see

Appendix B),

$$(2\pi\hbar)^n b(0) = \int_{\mathbb{R}^{2n}} F_\sigma W\psi(z)\rho_\sigma(-z)dz$$

$$= \int_{\mathbb{R}^{2n}} W\psi(z)\rho(z)dz$$

hence the inequality (14.10) since $b(0) \geq 0$. □

### 14.1.3. The quantum covariance matrix

We now address a difficult question which is closely related to the results of previous section: *When is a real symmetric $2n \times 2n$ matrix $\Sigma$ the covariance matrix of a mixed quantum state $\widehat{\rho}$?*

We will need the following crucial lemma, due to Narcowich [71]; its proof explicitly makes use of the quantum Bochner theorem:

**Lemma 181.** *Let $a : \mathbb{R}^{2n} \longrightarrow \mathbb{C}$ be continuous and twice continuously differentiable in a neighborhood of $0$. If $a$ is of $\hbar$-positive type, then we have*

$$-a''(0) + \frac{i\hbar}{2}J \geq 0, \tag{14.12}$$

*where $a'' = D^2 a$ is the Hessian matrix of $a$.*

**Proof.** We refer to [71, Lemma 2.1] for computational details. For $(\lambda_1, \ldots, \lambda_m) \in \mathbb{C}^m$ and $\varepsilon \in \mathbb{R}$ we define the function

$$R(\varepsilon) = \sum_{j,k=1}^m \overline{\lambda_j}\lambda_k e^{-\frac{i\varepsilon^2}{2\hbar}\sigma(z_j, z_k)} a(\varepsilon(z_j - z_k)).$$

If $a$ is of $\hbar$-positive type we have $R(\varepsilon) \geq 0$ for every $\varepsilon$; choose now the $\lambda_j$ such that $\sum_j \lambda_j = 0$; then $R(0) = 0$ and $R''(0) \geq 0$. An elementary but long calculation shows that we then have

$$R''(0) = (-2a''(0) + i\hbar^{-1}J)\zeta \cdot \zeta,$$

where $\zeta = \sum_j \lambda_j z_j \in \mathbb{C}^{2n}$. The $\lambda_j$ and $z_j$ being arbitrary numbers we thus have $-2a''(0) + i\hbar^{-1}J \geq 0$, proving the lemma. □

**Proposition 182.** *The covariance matrix $\Sigma_{\widehat{\rho}}$ of a density operator $\widehat{\rho}$ satisfies the quantum condition*

$$\Sigma_{\widehat{\rho}} + \frac{1}{2}i\hbar J \geq 0. \tag{14.13}$$

**Proof.** The matrix $\Sigma_{\widehat{\rho}} + \frac{1}{2}i\hbar J$ is Hermitian since $\Sigma_{\widehat{\rho}}$ is symmetric and the transposed of $J$ is $-J$. We next remark that $\Sigma_{\widehat{\rho}} = \Sigma_{\widehat{\rho}_0}$ where $\rho_0$ is defined by $\rho_0(z) = \rho(z + \langle z \rangle_{\rho})$ where $\langle z \rangle_{\widehat{\rho}}$ is the expectation value of $z = (x, p)$ in the state $\widehat{\rho}$: we have $\langle z \rangle_{\widehat{\rho}_0} = 0$ and hence

$$\text{Cov}(x_j, x_k)_{\widehat{\rho}_0} = \int_{\mathbb{R}^{2n}} x_j x_k \rho_0(z)dz = \text{Cov}(x_j, x_k)_{\widehat{\rho}};$$

similarly $\text{Cov}(x_j, x_k)_{\widehat{\rho}_0} = \text{Cov}(x_j, x_k)_{\widehat{\rho}}$ and $\text{Cov}(p_j, p_k)_{\widehat{\rho}_0} = \text{Cov}(p_j, p_k)_{\widehat{\rho}}$. It is thus sufficient to prove the proposition for the centered density operator $\widehat{\rho}_0$. A direct calculation shows that the Hessian matrix $\rho_{0,\sigma}''(0)$ satisfies

$$\hbar^2 \rho_{0,\sigma}''(0) = (2\pi\hbar)^{-n} \begin{pmatrix} -\Sigma_{pp,\rho_0} & \Sigma_{xp,\rho_0} \\ \Sigma_{px,\rho_0} & -\Sigma_{xx,\rho_0} \end{pmatrix}$$

and hence

$$\hbar^2 \rho_{0,\sigma}''(0) = \left(\frac{1}{2\pi\hbar}\right)^n J\Sigma_{\widehat{\rho}}J. \tag{14.14}$$

Since we have $\rho_\sigma = (2\pi\hbar)^{-n} a_\sigma$ the positivity of $\widehat{\rho}$ implies, taking condition (14.12) in Lemma 181 into account that

$$M = -2\hbar^{-1} J\Sigma_{\widehat{\rho}}J + iJ \geq 0.$$

The condition $M \geq 0$ being equivalent to $J^T M J \geq 0$; the inequality (14.13) follows.                                                                                    □

The following consequence of the condition (14.13) has already been proved in a particular case in Chapter 12.

**Corollary 183.** *Let $\widehat{\rho}$ be a density operator. Then the Robertson–Schrödinger inequalities*

$$(\Delta x_j)_{\widehat{\rho}}^2 (\Delta p_j)_{\widehat{\rho}}^2 \geq (\text{Cov}(x_j, p_j)_{\widehat{\rho}})^2 + \frac{1}{4}\hbar^2 \tag{14.15}$$

*hold.*

**Proof.** It follows from (14.13) by the same argument as in Proposition 155 of Chapter 12.                                                                                □

**Remark 184.** We emphasize again (cf. Chapter 13) that the inequalities (14.15), which are a consequence of the quantum condition (14.13), does not imply that $\Sigma_{\widehat{\rho}}$ is a quantum covariance matrix.

An interesting application of the KLM conditions is the case of Gaussian mixed states. It turns out that for such states condition (14.13) is not only necessary, but also sufficient.

**Proposition 185.** *Let $\rho_\Sigma$ be the real Gaussian function defined by*

$$\rho_\Sigma(z) = \frac{1}{(2\pi)^n \sqrt{\det \Sigma}} e^{-\frac{1}{2}\Sigma^{-1} z \cdot z}, \tag{14.16}$$

*where $\Sigma$ is a positive definite symmetric real $2n \times 2n$ matrix. The operator $\widehat{\rho}_\Sigma = (2\pi\hbar)^n \mathrm{Op_W}(\rho_\Sigma)$ is a density operator if and only if the symplectic eigenvalues $\lambda_j$ of $\Sigma$ are such that $\lambda_j \geq \frac{1}{2}\hbar(1 \leq j \leq n)$; equivalently*

$$\Sigma + \frac{i\hbar}{2}J \geq 0. \tag{14.17}$$

*When this condition is satisfied we have $\Sigma = \Sigma_{\widehat{\rho}}$.*

**Proof.** We begin by noting that the function $\rho_\Sigma$ is a classical probability density: we have $\rho_\Sigma > 0$ and

$$\int_{\mathbb{R}^{2n}} \rho_\Sigma(z)dz = 1.$$

The necessity of condition (14.17) easily follows from Proposition 182; let us however give an independent proof. Suppose that $\rho_\Sigma$ is the Wigner transform of a density operator; then

$$\rho_\Sigma(z) = \sum_j \alpha_j W\psi_j(z)$$

for functions $\psi_j \in L^2(\mathbb{R}^n)$, the $\alpha_j > 0$ summing up to one. It is no restriction, using a symplectic diagonalization $\Sigma = S^T D S$ (see Appendix C) and replacing each $\psi_j$ into $\widehat{S}^{-1}\psi_j$, to assume that $\Sigma$ is diagonal, of the type

$$\Sigma = D = \begin{pmatrix} \Lambda & 0 \\ 0 & \Lambda \end{pmatrix},$$

where $D$ is the diagonal matrix

$$D = \begin{pmatrix} \lambda_1 & 0 & \cdots & 0 \\ 0 & \lambda_2 & \cdots & 0 \\ \vdots & \vdots & \ddots & \vdots \\ 0 & 0 & \cdots & \lambda_n \end{pmatrix},$$

where the $\lambda_j$ are the symplectic eigenvalues of $\Sigma$. Assuming in addition that $\psi_j \in L^1(\mathbb{R}^n)$ the marginal conditions satisfied by the Wigner transform imply that

$$\int_{\mathbb{R}^n} \rho_\Sigma(z)dp = \sum_j \alpha_j \int_{\mathbb{R}^n} W\psi_j(z)dp = \sum_j \alpha_j |\psi_j(x)|^2,$$

$$\int_{\mathbb{R}^n} \rho_\Sigma(z)dx = \sum_j \alpha_j \int_{\mathbb{R}^n} W\psi_j(z)dxp = \sum_j \alpha_j |F\psi_j(p)|^2.$$

Since we have

$$\int_{\mathbb{R}^n} \rho_\Sigma(z)dp = \frac{1}{(2\pi)^{n/2}\sqrt{\det\Lambda}} e^{-\frac{1}{2}\Lambda^{-1}x\cdot x},$$

$$\int_{\mathbb{R}^n} \rho_\Sigma(z)dx = \frac{1}{(2\pi)^{n/2}\sqrt{\det\Lambda}} e^{-\frac{1}{2}\Lambda^{-1}p\cdot p},$$

it follows that

$$\sum_j \alpha_j |\psi_j(x)|^2 = \frac{1}{(2\pi)^{n/2}\sqrt{\det\Lambda}} e^{-\frac{1}{2}\Lambda^{-1}x\cdot x},$$

$$\sum_j \alpha_j |F\psi_j(p)|^2 = \frac{1}{(2\pi)^{n/2}\sqrt{\det\Lambda}} e^{-\frac{1}{2}\Lambda^{-1}p\cdot p},$$

hence there exist constants $C_j$ such that

$$|\psi_j(x)| \leq C_j e^{-\frac{1}{2}\Lambda^{-1}x\cdot x}, \quad |F\psi_j(p)| \leq C_j e^{-\frac{1}{2}\Lambda^{-1}p\cdot p}.$$

In view of the Hardy's uncertainty principle studied in Chapter 10 this requires that we have $\hbar^2\lambda_j^{-2}/4 \leq 1$, that is $\lambda_j \geq \hbar/2$ (which is equivalent to the condition (14.17)). Let us next prove the sufficiency of the condition (14.17). A straightforward computation of Gaussians shows that the reduced symplectic transform $F_\Diamond \rho_\Sigma = (\rho_\Sigma)_\Diamond$ is given by

$$(\rho_\Sigma)_\Diamond(z) = e^{\frac{1}{2}J\Sigma Jz\cdot z} = e^{-\frac{1}{2}\Sigma Jz\cdot Jz}.$$

We have $(\rho_\Sigma)_\Diamond(0) = 1$ and an easy calculation shows that the matrix $\Lambda = (\Lambda_{jk})_{1\leq j,k\leq N}$ in Definition 174 is positive semidefinite if and only if the matrix $M$ with entries

$$M_{jk}(z_j, z_k) = e^{\frac{1}{2}(-J\Sigma J+i\hbar J)z_j\cdot z_k}$$

is positive semidefinite; this in turn is equivalent to condition (14.17) (we refer to Narcowich [69] and Narcowich and O'Connell [73] for the computational details). □

**Remark 186.** Suppose that all the symplectic eigenvalues of $\Sigma$ are equal to $\frac{1}{2}\hbar$. Then $\Sigma = S^T DS = S^T S$ for some $S \in \mathrm{Sp}(n)$ and $\rho$ is the Wigner transform of a generalized Gaussian function $\psi_M$ as considered in Chapter 9. In all other cases (i.e. as soon there exists $j$ such that $\lambda_j < \frac{1}{2}\hbar$) the function $\rho$ is the Wigner transform of the density operator of a mixed state, i.e. the purity $\mathrm{Tr}(\widehat{\rho}^2)$ is $> 1$.

## 14.2. The Narcowich–Wigner Spectrum

Let us generalize the notion of $\hbar$-positivity.

### 14.2.1. $\eta$-Positive functions

The following definition is due to Bröcker and Werner [13]; they use the terminology "Wigner spectrum" which we prefer to call "Narcowich–Wigner spectrum" to give credit to the work of Narcowich [69–71] where the notion appeared explicitly for the first time.

**Definition 187.** Let $f \in \mathcal{S}'(\mathbb{R}^{2n})$ and $\eta \in \mathbb{R}$. We say that $f$ is $\eta$-positive if for any choice of $z_1, \ldots, z_N$ in $\mathbb{R}^{2n}$ the (complex) $N \times N$ matrix $\Lambda = (\Lambda_{jk}(z_j, z_k))_{1 \le j,k \le N}$ defined by

$$\Lambda_{jk}(z_j, z_k) = e^{-\frac{i\eta}{2}\sigma(z_j, z_k)} a_{\Diamond}(z_j - z_k) \tag{14.18}$$

is positive semidefinite.

Suppose that $f$ is the reduced symplectic transform $a_{\Diamond}$ of a symbol $a \in \mathcal{S}'(\mathbb{R}^{2n})$. Then we can define the Narcowich–Wigner spectrum of $a$.

**Definition 188.** Let $a \in \mathcal{S}'(\mathbb{R}^{2n})$. The set of values of $\eta$ for which $a_{\Diamond}$ is $\eta$-positive is called the *Narcowich–Wigner spectrum of $a$* and is denoted by $\mathrm{NWS}(a)$.

Alternatively, $\mathrm{NWS}(f)$ is thus the set of all real numbers $\eta$ such that for every integer $N \ge 1$ and every sequence $(z_1, \ldots, z_N)$ we have

$$\sum_{1 \le j,k \le N} \lambda_j \overline{\lambda_k} e^{-\frac{i\eta}{2}\sigma(z_j, z_k)} a_{\Diamond}(z_j - z_k) \ge 0 \tag{14.19}$$

$\lambda = (\lambda_1, \ldots, \lambda_N) \in \mathbb{C}^N$. It should be emphasized that in the verification of the condition above it is assumed that $a_{\Diamond}$ does not depend explicitly on the variable $\eta$.

Functions of $\eta$-positive type form a cone: if $f$ and $g$ are of $\eta$-positive type then so is $\lambda f + \mu g$ for all $\lambda \ge 0$ and $\mu \ge 0$.

Notice that if we choose $\eta = 0$ we recover the usual definition of a function of positive type discussed above in relation with Bochner's theorem, replacing the characteristic function with the symplectic Fourier transform: the function $a$ is of positive type if for each integer $N$ the $N \times N$ matrix with entries $a_\Diamond(z_j - z_k)$ is positive semidefinite.

The proof of the following properties, which are of an elementary nature, can be found in the paper by Bröcker and Werner [13].

**Proposition 189.** *Let $a$ and $b$ be tempered distributions on $\mathbb{R}^{2n}$. Then we have the following proposition.*

(i)   *The Narcowich–Wigner spectrum* $\mathrm{NWS}(a)$ *is a closed subset of* $\mathbb{R}$.
(ii)  *If $a$ is continuous and non-zero, then* $\mathrm{NWS}(a)$ *is bounded.*
(iii) *The set* $\mathrm{NWS}(a)$ *is symmetric:* $\eta \in \mathrm{NWS}(a)$ *if and only if* $-\eta \in \mathrm{NWS}(a)$.
(iv)  *Let $\lambda \in \mathbb{R}$ be non-zero and set $a_\lambda(x) = a(\lambda x)$; we have* $\mathrm{NWS}(a_\lambda) = \lambda^2 \mathrm{NWS}(a)$.
(v)   *We have* $\mathrm{NWS}(a) + \mathrm{NWS}(b) \subset \mathrm{NWS}(ab)$.

That $\mathrm{NWS}(f)$ is closed can be seen directly as follows: let $(\eta_m)_m$ be a sequence of real numbers in $\mathrm{NWS}(f)$ such that $\lim_{m \to \infty} \eta_m = \eta$. Then the matrices $\Lambda_m = (\Lambda_{m,jk}(z_j, z_k))$ formed according to (14.18) replacing $\eta$ with $\eta_m$ are positive semidefinite for every $m$; since the limit matrix $\Lambda = \lim_{m \to \infty} \Lambda_m$ is also positive semidefinite we have $\eta \in \mathrm{NWS}(f)$.

### 14.2.2.  The Narcowich–Wigner spectrum of some states

Let $\eta \neq 0$ be a real parameter as above. We replace in Wigner–Weyl–Moyal theory Planck's constant $\hbar$ with this parameter. Technically, this amounts to redefine the Heisenberg–Weyl and Grossmann–Royer operators as

$$\widehat{T}_\eta(z_0)\psi(x) = e^{\frac{i}{\eta}(p_0 \cdot x - \frac{1}{2}p_0 \cdot x_0)}\psi(x - x_0),$$

$$\widehat{R}_\eta(z_0)\psi(x) = e^{\frac{2i}{\eta}p_0 \cdot (x - x_0)}\psi(2x_0 - x).$$

For instance, the usual Wigner transform $W\psi$ of a function $\psi \in L^2(\mathbb{R}^n)$ is then replaced with the $\eta$-Wigner transform $W_\eta\psi$ defined by

$$W_\eta\psi(z) = \left(\frac{1}{2\pi\eta}\right)^n \int_{\mathbb{R}^n} e^{-\frac{i}{\eta}p \cdot y}\psi\left(x + \frac{1}{2}y\right)\overline{\psi\left(x - \frac{1}{2}y\right)}dy \qquad (14.20)$$

and the Weyl operator $\widehat{A} = \mathrm{Op}_{\mathrm{W}}(a)$ becomes the $\eta$-Weyl operator $\widehat{A}_\eta = \mathrm{Op}_{\mathrm{W},\eta}(a)$ formally defined by the pairing

$$\langle \widehat{A}_\eta \psi, \overline{\phi} \rangle = \langle a, W_\eta(\psi, \phi) \rangle.$$

Consider now a bounded self-adjoint trace one $\eta$-Weyl operator $\widehat{\rho}_\eta$ on $L^2(\mathbb{R}^n)$. We will call $\widehat{\rho}_\eta$ an $\eta$-density operator, it; is in addition positive semidefinite: $\widehat{\rho}_\eta \geq 0$ (when $\eta = \hbar$ we thus recover the usual notion of density operator). Equivalently, the Weyl symbol of $\widehat{\rho}_\eta$ is $(2\pi\eta)^n \rho_\eta$ where $\rho_\eta$ is a convex sum of $\eta$-Wigner transforms

$$\rho_\eta(z) = \sum_j \alpha_j W_\eta \psi_j.$$

We will call the set of values $\eta$ for which $\widehat{\rho}_\eta \geq 0$ the Narcowich–Wigner spectrum of $\widehat{\rho}_\eta$; it is thus the set $\mathrm{NWS}(\rho_\eta)$. Now, very little is known about the Narcowich–Wigner spectrum for general operators $\widehat{\rho}_\eta$, except when $\rho_\eta$ is a Gaussian. In this case the answer is easily obtained by reversing the proof of Proposition 185.

**Proposition 190.** *Let $\rho_\Sigma$ be the normalized Gaussian* (14.16). *We have*

$$\mathrm{NWS}(\rho_\Sigma) = [-2\lambda_{\min}, 2\lambda_{\min}],$$

*where $\lambda_{\min}$ is the smallest symplectic eigenvalue of the covariance matrix $\Sigma$.*

**Proof.** Assume that $\eta > 0$. Replacing everywhere $\hbar$ with $\eta$ in the proof of Proposition 185 we see that $\widehat{\rho}_\eta$ is an $\eta$-density operator if and only if the symplectic eigenvalues $\lambda_j$ of $\Sigma$ satisfy the inequalities $\lambda_j \geq \frac{1}{2}\eta$ $(1 \leq j \leq n)$, that is, equivalently $\lambda_{\min} \geq \frac{1}{2}\eta$, or $\eta \in (0, \lambda_{\min}]$. The case $\eta \leq 0$ easily follows: it is clear that 0 belongs to the Narcowich–Wigner spectrum since $\widehat{\rho}_{\eta=0}$ is just a classical normal distribution, and the case $\eta < 0$ follows because $\eta$ is in the Narcowich–Wigner spectrum if and only if $-\eta$ is. $\quad\square$

The result above can be stated in the following equivalent way.

**Corollary 191.** *Let $F$ be a positive-definite real symmetric $2n \times 2n$ matrix and set*

$$\rho(z) = \left(\frac{1}{\pi\hbar}\right)^n (\det F) e^{-\frac{1}{\hbar} Fz \cdot z}. \tag{14.21}$$

*The corresponding operator $\widehat{\rho}$ is a density operator if and only if all the symplectic eigenvalues of $F$ are $\leq \hbar$ and we have*

$$\mathrm{NWS}(\rho) = [-\hbar, \hbar]. \tag{14.22}$$

Notice that $\hat{\rho}$ is the density operator of a pure state if and only if all the symplectic eigenvalues of $F$ are equal to one, in which case $F = S^T S$ for some $S \in \mathrm{Sp}(n)$ and $\rho$ is the Wigner transform of a generalized Gaussian (= squeezed coherent state).

**Example 192.** In the case $n = 1$ consider the Gaussian function

$$\rho_\Sigma = \frac{1}{2\pi\sigma_X\sigma_P} \exp\left[-\frac{1}{2}\left(\frac{x^2}{\sigma_X^2} + \frac{p^2}{\sigma_P^2}\right)\right].$$

We have here $\lambda_{\min} = \sigma_X\sigma_P$; in view of the uncertainty principle we must have $\sigma_X\sigma_P \geq \frac{1}{2}\hbar$ if $\hat{\rho}_\Sigma$ represents a mixed quantum state, that is

$$\mathrm{NWS}(\rho_\Sigma) = [-2\lambda_{\min}, 2\lambda_{\min}] = [-\hbar, \hbar].$$

In the non-Gaussian case, the situation is quite intricate (see the examples in [13]). Dias and Prata [19] have also obtained interesting results. A complete description of the Narcowich–Wigner spectrum for arbitrary mixed states is however lacking altogether. This is certainly an interesting topic of research since it could lead to a better understanding of the sensitivity of mixed quantum states to small variations of Planck's constant; for instance in the Gaussian case the equality (14.22) shows that the operator $\hat{\rho}$ corresponding to (14.21) remains a density operator when one decreases the value of Planck's constant but not if one increases it; in the latter the mixed quantum state turns into a classical state.

**Main Formulas in Chapter 14**

Reduced symplectic FT        $F_\Diamond a(z) = \left(\frac{1}{2\pi}\right)^n \int_{\mathbb{R}^{2n}} e^{i\sigma(z,z')} a(z') dz'$

Bochner conditions           $\sum_{1 \leq j,k \leq N} \lambda_j \overline{\lambda_k} a_\Diamond(z_j - z_k) \geq 0$

KLM conditions               $\sum_{1 \leq j,k \leq N} \lambda_j \overline{\lambda_k} e^{-\frac{i\hbar}{2}\sigma(z_j,z_k)} f(z_j - z_k) \geq 0$

Condition for $\hat{\rho} \geq 0$        $\int_{\mathbb{R}^{2n}} \rho(z) W\psi(z) dz \geq 0$

Quantum condition            $\Sigma_{\hat{\rho}} + \frac{1}{2}i\hbar J \geq 0$

Robertson–Schrödinger UP     $(\Delta x_j)_{\hat{\rho}}^2 (\Delta p_j)_{\hat{\rho}}^2 \geq (Cov(x_j, p_j)_{\hat{\rho}})^2 + \frac{1}{4}\hbar^2$

Gaussian distribution        $\rho_\Sigma(z) = \frac{1}{(2\pi)^n\sqrt{\det\Sigma}} e^{-\frac{1}{2}\Sigma^{-1}z\cdot z}$

Chapter 15

# Wigner Transform
# and Quantum Blobs

**Summary 193.** Quantum blobs are a geometric picture of the Wigner transform of generalized Gaussians (= squeezed coherent states). They allow a symplectically invariant coarse-graining of phase space. Quantum blobs can also be described using the symplectic capacity of phase space ellipsoids.

The "quantum blobs" we introduce in this chapter lead to a coarse-graining of quantum phase space compatible with the strong uncertainty principle studied in Chapter 12. They are the images by (linear) symplectic automorphisms of phase space balls with radius $\sqrt{\hbar}$ and give a geometric description of the level sets of the Wigner transform of Gaussians (the "covariance ellipsoid"). Quantum blobs were introduced and studied in our papers [31, 32, 38, 40] and further developed in our book [33]; also see [17, 26, 46].

## 15.1. Quantum Blobs and Phase Space

Let $\psi_M$ be a generalized Gaussian (formula (9.1) in Chapter 9). Its Wigner transform is a phase space Gaussian

$$W\psi_M(z) = \left(\frac{1}{\pi\hbar}\right)^n e^{-\frac{1}{\hbar}S^T Sz \cdot z}. \tag{15.1}$$

The equation $S^T Sz \cdot z \leq \hbar$ determines an ellipsoid in $\mathbb{R}^{2n}$. That ellipsoid is a *quantum blob*.

### 15.1.1. Geometric definition of a quantum blob

In what follows all the symplectic capacities are supposed to be linear (for extensions to the case of general symplectomorphisms see [46]).

Recall (see Appendix A) that the affine symplectic group is the semi-direct product $\mathrm{ASp}(n) = \mathrm{Sp}(n) \ltimes T(n)$; it is the group generated by the symplectic group $\mathrm{Sp}(n)$ and the phase space translations $T(z_0) : z \longmapsto z + z_0$. A typical element of $\mathrm{ASp}(n)$ can be written $ST(z)$ or $T(z)S$ and we have the covariance formulas

$$ST(z) = T(Sz)S, \quad T(z)S = ST(S^{-1}z). \tag{15.2}$$

**Definition 194.** A quantum blob is the image $\mathcal{B}$ of the ball $B^{2n}(\sqrt{\hbar})$ by an element of the affine symplectic group $\mathrm{ASp}(n)$.

Quantum blobs are thus phase space ellipsoids. A quantum blob has symplectic capacity $\pi\hbar$ as follows from the equality

$$c(B^{2n}(\sqrt{\hbar})) = \pi\hbar = \frac{1}{2}h$$

and the symplectic invariance of symplectic capacities.

Any subset $\Omega$ of $\mathbb{R}^n$ such that $c(\Omega) \geq \frac{1}{2}h$ for some symplectic capacity $c$ contains a quantum blob $\mathcal{B}$: by definition of the Gromov width $c_{\mathrm{G}}$ there exists a symplectic transformation $f$ such that

$$\mathcal{B} = f(B^{2n}(\sqrt{\hbar})) \subset \Omega.$$

Suppose conversely that $c_{\mathrm{G}}(\Omega) \geq \frac{1}{2}h$ then there exists $f \in \mathrm{ASp}(n)$ such that $f(B^{2n}(\sqrt{\hbar})) \subset \Omega$ and hence $\Omega$ contains a quantum blob.

**Proposition 195.** *Let $\Omega$ be a phase space ellipsoid. Assume that $c(\Omega) = \frac{1}{2}h$. Then $\Omega$ contains a unique quantum blob $\mathcal{B}$.*

**Proof.** Since all symplectic capacities agree on ellipsoids, the condition $c(\Omega) = \frac{1}{2}h$ is equivalent to $c_{\mathrm{G}}(\Omega) = \frac{1}{2}h$. Translating if necessary $\Omega$ it is sufficient to prove the claim in the case where all the considered ellipsoids are centered at zero and the elements of $\mathrm{ASp}(n)$ are linear (i.e. belong to $\mathrm{Sp}(n)$). The assumption $c_{\mathrm{G}}(\Omega) = \frac{1}{2}h$ thus means that $B^{2n}(\sqrt{\hbar})$ is the largest ball $B^{2n}(R)$ such that $S(B^{2n}(\sqrt{\hbar})) \subset \Omega$ for some $S \in \mathrm{Sp}(n)$. Thus, $\mathcal{B} = S(B^{2n}(\sqrt{\hbar}))$ is a quantum blob contained in $\Omega$. Assume now that there exists $S' \in \mathrm{Sp}(n)$ such that $\mathcal{B}' = S'(B^{2n}(\sqrt{\hbar}))$ is also contained in the ellipsoid $\Omega$ and let us show that $\mathcal{B}' = \mathcal{B}$; the uniqueness will follow.

Conjugating if necessary $S$ and $S'$ with an adequately chosen symplectic matrix we may assume, using Williamson's diagonalization theorem, that

$$\Omega : \sum_{j=1}^{n} \lambda_j (x_j^2 + p_j^2) \le 1,$$

where the $\lambda_j$ are the symplectic eigenvalues of the matrix of $\Omega$. The condition $c(\Omega) = \frac{1}{2}h$ means that $\lambda_1 = 1$. Setting $U = S'S^{-1}$ we have $S' = US$. Let us show that $U \in U(n)$; the proposition will follows since then $U(B^{2n}(\sqrt{\hbar})) = B^{2n}(\sqrt{\hbar})$ so that $S(B^{2n}(\sqrt{\hbar})) = S'(B^{2n}(\sqrt{\hbar}))$. Clearly $U \in \mathrm{Sp}(n)$ so it suffices, in view of the identity (A.11), to show that $U$ is in addition a rotation; for this it is sufficient to check that $UJ = JU$. Set $R = D^{1/2} U D^{-1/2}$ where $D = \mathrm{diag}(\lambda_1, \ldots, \lambda_n)$; we have $U^T D U = D$ hence $R^T R = I$ so that $R$ is orthogonal. Let us prove that $R$ is in addition symplectic. Since $J$ commutes with every power of the diagonal matrix $D$ we have, taking into account the relation $JU = (U^T)^{-1}J$ (because $U$ is symplectic):

$$JR = D^{1/2} JU D^{-1/2} = D^{1/2}(U^T)^{-1} J D^{-1/2}$$
$$= D^{1/2}(U^T)^{-1} D^{-1/2} J = (R^T)^{-1} J$$

hence $R^T JR = J$ so that $R$ is indeed symplectic. Since $R$ is also a rotation we have $R \in U(n)$ and thus $JR = RJ$. Since $U = D^{-1/2} R D^{1/2}$ we have

$$JU = JD^{-1/2} R D^{1/2} = D^{-1/2} JR D^{1/2}$$
$$= D^{-1/2} RJ D^{1/2} = D^{-1/2} R D^{1/2} J = UJ$$

so that $U \in U(n)$ as claimed. □

### 15.1.2. Quantum phase space

Let us denote by $\mathrm{QPS}(2n)$ the set of all linear quantum blobs in $\mathbb{R}^{2n}$. It is very easy to equip $\mathrm{QPS}(2n)$ with a natural topology. Consider first the case of a quantum blob $\mathcal{B} = S(B^{2n}(\sqrt{\hbar}))$ centered at the origin; $\mathcal{B}$ is a centered ellipsoid uniquely represented by the inequality $(S^T S)z \cdot z \le \hbar$. The symplectic matrix $S^T S$ is positive-definite and, conversely, every positive-definite symplectic matrix can be written in the form $S^T S$ in view of the polar decomposition theorem. There is thus a bijective correspondence between the set $\mathrm{QPS}_0(2n)$ of entered linear quantum blobs and positive definite symmetric matrices. These matrices form a topological manifold which is homeomorphic to the coset space $\text{Sp}(n)/U(n)$:

$$\text{QPS}_0(2n) = \text{Sp}(n)/U(n).$$

This immediately follows from the symplectic polar decomposition theorem. As a manifold $\text{Sp}(n)/U(n)$ has dimension,

$$\dim(\text{Sp}(n)/U(n)) = \dim \text{Sp}(n) - \dim U(n)$$
$$= n(2n + 1) - n^2$$
$$= n(n + 1).$$

We can thus identify the set of all linear quantum blobs centered at the origin with the Euclidean space $\mathbb{R}^{n(n+1)}$. Now, an arbitrary linear quantum blob is obtained from a centered quantum blob by a phase space translation; such translations form a group isomorphic to $\mathbb{R}^{2n}$. It follows, by dimension count, that we have the identification

$$\text{QPS}(2n) \equiv \mathbb{R}^{n(n+1)} \times \mathbb{R}^{2n} \equiv \mathbb{R}^{n(n+3)}.$$

**Remark 196.** Since symplectomorphisms are volume preserving, the volume of a quantum blob is

$$\text{Vol}(\mathcal{B}) = \text{Vol}(B^{2n}(\sqrt{\hbar})) = \frac{1}{n!}\left(\frac{h}{2}\right)^n;$$

their volume is thus much smaller than that of a traditional quantum cell (which is $h^n$), and decreases very quickly with $n$.

## 15.2. Quantum Blobs and the Wigner Transform

It turns out that there is a very interesting a subtle relation between quantum blobs and the Wigner transforms of Gaussians.

### 15.2.1. The basic example

Recall from Chapter 9 that the Wigner transform of a generalized Gaussian

$$\psi_M(x) = \left(\frac{1}{\pi\hbar}\right)^{n/4} (\det X)^{1/4} e^{-\frac{1}{2\hbar}Mx^2},$$

where $M = X + iY$, $X = X^T > 0$, $Y = Y^T$, is the phase space Gaussian

$$W\psi_M(z) = \left(\frac{1}{\pi\hbar}\right)^n e^{-\frac{1}{\hbar}Gz^2}, \qquad (15.3)$$

where $G$ is the symmetric symplectic matrix $G = S^T S$ where

$$S = \begin{pmatrix} X^{1/2} & 0 \\ X^{-1/2}Y & X^{-1/2} \end{pmatrix}. \tag{15.4}$$

If we now view $W\psi_M(z)$ as a probability distribution, it is easy to see that the corresponding covariance matrix is given by

$$\Sigma = \frac{\hbar}{2}G^{-1}. \tag{15.5}$$

In fact, inserting $M = \frac{\hbar}{2}\Sigma^{-1}$ in (15.3) we get

$$W\psi_M(z) = \left(\frac{1}{\pi\hbar}\right)^n e^{-\frac{1}{2}\Sigma^{-1}z^2}.$$

Since $\det G = 1$ (because $G$ is symplectic), this shows that $W\psi_M$ is a normal probability distribution with covariance matrix $\Sigma$.

**Proposition 197.** *Let $\Sigma$ be the covariance matrix determined by the generalized Gaussian $\psi_M$ and denote by $\Omega_\Sigma$ the associated covariance ellipsoid:*

$$\Omega_\Sigma : \frac{1}{2}\Sigma^{-1}z \cdot z \leq 1.$$

*This ellipsoid is a quantum blob, hence, in particular $c(\Omega_\Sigma) = \pi\hbar$.*

**Proof.** In view of (15.5) the inequality $Gz^2 \leq \hbar$ is equivalent to $\frac{1}{2}\Sigma^{-1}z \cdot z \leq 1$ hence the result. $\square$

### 15.2.2. Covariance ellipsoid and quantum blobs

Let us introduce the following terminology.

**Definition 198.** Let $\Sigma$ be a covariance matrix. The phase space ellipsoid

$$\Omega_\Sigma = \left\{ z \in \mathbb{R}^{2n} : \frac{1}{2}\Sigma^{-1}z \cdot z \leq 1 \right\}$$

is called the "covariance ellipsoid" associated with $\Sigma$.

The following geometric result reveals the deep connection between the (strong) uncertainty principle and the notion of quantum blob introduced in Chapter 15.

**Proposition 199.** *A covariance matrix* $\Sigma$ *is a quantum covariance matrix, that is*

$$\Sigma + \frac{i\hbar}{2}J \geq 0 \tag{15.6}$$

*if and only if the associated covariance ellipsoid* $\Omega_\Sigma$ *has symplectic capacity at least* $\pi\hbar$

$$c(\Omega_\Sigma) \geq \pi\hbar; \tag{15.7}$$

*equivalently,* $\Omega_\Sigma$ *contains a quantum blob.*

**Proof.** The condition (15.7) is equivalent to say that the covariance ellipsoid contains a quantum blob in view of Proposition 195. In Appendix D it is shown that the symplectic capacity (linear, or not) of the phase space ellipsoid

$$\Omega_M : Mz \cdot z \leq 1 \tag{15.8}$$

is given by the formula

$$c(\Omega_M) = \frac{\pi}{\lambda_{\max}}, \tag{15.9}$$

where $\lambda_{\max}$ is the largest symplectic eigenvalue of $M$ (recall from Appendix C that the symplectic eigenvalues of a symmetric positive definite real $2n \times 2n$ matrix are the positive numbers $\lambda_1, \lambda_2, \ldots, \lambda_n$ defined as follows: $\pm i\lambda_1, \pm i\lambda_2, \ldots, \pm i\lambda_n$ are the eigenvalues of the matrix $JM$ where $J$ is the standard symplectic matrix). The covariance ellipsoid $\Omega_\Sigma$ corresponds to the choice $M = \frac{1}{2}\Sigma^{-1}$, that is $\Omega_\Sigma = \Omega_{\frac{1}{2}\Sigma^{-1}}$ and the symplectic eigenvalues $\lambda_1, \lambda_2, \ldots, \lambda_n$ of $M$ are thus $\frac{1}{2}\mu_1^{-1}, \frac{1}{2}\mu_2^{-1}, \ldots, \frac{1}{2}\mu_n^{-1}$ where $\mu_1, \mu_2, \ldots, \mu_n$ are the symplectic eigenvalues of the covariance matrix $\Sigma$. Hence formula (15.9) is equivalent to

$$c(\Omega_\Sigma) = 2\pi\mu_{\min},$$

where $\mu_{\min} = 1/2\lambda_{\max}$ is the smallest symplectic eigenvalue of $\Sigma$. Let us show that $\mu_{\min} \geq \frac{1}{2}\hbar$, the inequality (15.7) will follow. The condition (15.6) is equivalent to $M^{-1} + i\hbar J \geq 0$; using a Williamson diagonalization of $M$ (Appendix C) this is in turn equivalent to $D^{-1} + i\hbar J \geq 0$ where

$$D = \begin{pmatrix} \Lambda & 0 \\ 0 & \Lambda \end{pmatrix}, \quad \Lambda = \begin{pmatrix} \lambda_1 & 0 & \cdots & 0 \\ 0 & \lambda_2 & \cdots & 0 \\ \vdots & \vdots & & \vdots \\ 0 & 0 & \cdots & \lambda_n \end{pmatrix}.$$

The characteristic polynomial of $D^{-1} + i\hbar J$ is easily calculated; it is the product $P(t) = P_1(t) \cdots P_n(t)$ where

$$P_j(t) = t^2 - \lambda_j^{-1} t + \frac{1}{4} \lambda_j^{-2} - \frac{1}{4} \hbar^2.$$

The roots of $P(t)$ are the eigenvalues of $D^{-1} + i\hbar J$; they the real numbers $\frac{1}{2}(\lambda_j^{-1} \pm \hbar)$ hence $D^{-1} + i\hbar J \geq 0$ if and only if $\lambda_j^{-1} - \hbar \geq 0$ for every $j$, that is if and only if $\lambda_j^{-1} \geq \hbar$ for every $j$; this is equivalent to $\mu_j \geq \frac{1}{2} \hbar$ for all $j$, that is to $\mu_{\min} \geq \frac{1}{2} \hbar$ which we set out to prove. Reversing the argument, the condition is also sufficient. $\quad\square$

**Remark 200.** Let $\Sigma$ be a quantum covariance matrix and $\Omega_\Sigma$ the associated covariance ellipsoid. The result above shows that the condition $c(\Omega_\Sigma) \geq \frac{1}{2} h$ is equivalent to the Robertson–Schrödinger uncertainty inequalities

$$(\Delta x_j)^2 (\Delta p_j)^2 \geq \Delta(x_j, p_j)^2 + \frac{1}{4} \hbar^2.$$

The condition $c(\Omega_\Sigma) \geq \pi\hbar$ can thus be seen as a geometric formulation of the quantum uncertainty principle (it would be more accurate to speak about quantum indeterminacy).

Recall from Chapter 14 (Proposition 185) that a Gaussian function

$$\rho(z) = \frac{1}{(2\pi)^n \det \Sigma} e^{-\frac{1}{2}\Sigma^{-1} z \cdot z}$$

is the Wigner transform of a density operator if and only if the symplectic eigenvalues $\lambda_j$ of $\Sigma$ are such that $\lambda_j \geq \frac{1}{2} \hbar$ for $1 \leq j \leq n$; equivalently the inequality (15.6) holds. It immediately follows from the discussion above that:

**Proposition 201.** *A real Gaussian function*

$$\rho(z) = \frac{1}{(2\pi)^n \det \Sigma} e^{-\frac{1}{2}\Sigma^{-1} z \cdot z}$$

*with* $\Sigma = \Sigma^T > 0$ *is the Wigner transform of a density operator if and only if the covariance ellipsoid* $\Omega_\Sigma$ *contains a quantum blob.*

## 15.3. From One Quantum Blob to Another

Generalized Gaussians can be obtained from one another by simple metaplectic transformations only involving multiplication by "chirps" and unitary rescaling; similarly quantum blobs are obtained from each other by using the corresponding symplectic transformations.

### 15.3.1. The general case

We are following here the approach in de Gosson and Luef [46]. We begin by stating a technical result which gives a particular factorization result for symplectic matrices.

**Lemma 202.** *Let* $S = \begin{pmatrix} A & B \\ C & D \end{pmatrix}$ *be a symplectic matrix. We have*

$$S = \begin{pmatrix} A_0 & 0 \\ C_0 & A_0^{-1} \end{pmatrix} \begin{pmatrix} X_0 & -Y_0 \\ Y_0 & X_0 \end{pmatrix},$$

*where*

$$A_0 = (AA^T + BB^T)^{1/2} = A_0^T,$$
$$C_0 = (CA^T + DB^T)(AA^T + BB^T)^{-1/2},$$
$$X_0 + iY_0 = (AA^T + BB^T)^{-1/2}(A - iB).$$

**Proof.** It is purely computational; see [33, Proposition 2.29, p. 42 and Corollary 2.30, p. 43].                    □

A consequence of this lemma is as follows.

**Proposition 203.** *Every quantum blob* $\mathcal{B} = S(B^{2n}(\sqrt{\hbar}))$ *centered at the origin can be written* $\mathcal{B} = S_0(B^{2n}(\sqrt{\hbar}))$ *where the symplectic matrix* $S_0$ *is given by*

$$S_0 = \begin{pmatrix} I & 0 \\ P & I \end{pmatrix} \begin{pmatrix} L^{-1} & 0 \\ 0 & L \end{pmatrix} = V_{-P}M_L, \qquad (15.10)$$

*where* $L$ *and* $P$ *are symmetric (and* $L$ *invertible). In fact, if* $S = \begin{pmatrix} A & B \\ C & D \end{pmatrix}$ *then*

$$L = (AA^T + BB^T)^{-1/2}, \quad P = (CA^T + DB^T)(AA^T + BB^T)^{-1}. \quad (15.11)$$

**Proof.** It is a straightforward consequence of Lemma 202 since the symplectic matrix

$$U = \begin{pmatrix} X_0 & -Y_0 \\ Y_0 & X_0 \end{pmatrix}$$

is a rotation (hence $U \in U(n)$) and thus leaves $B^{2n}(\sqrt{\hbar})$ invariant since it is centered at the origin.                    □

The results above enable us to prove the main result of this section.

**Proposition 204.** *Let $\mathcal{B}$ be a linear quantum blob centered at $z_0$:*

$$\mathcal{B} = S_0(B^{2n}(z_0, \sqrt{\hbar}))$$

*where the symplectic matrix $S_0$ is given by (15.10) with $P$ and $L$ symmetric.*

(i) *The function*

$$W_{\mathcal{B}}(z) = (\pi\hbar)^{-n} e^{-\frac{1}{\hbar}G(z-z_0)^2}, \quad G = (S_0^T)^{-1}S_0^{-1}$$

*is independent of the choice of $S_0$ and is thus uniquely determined by the quantum blob $\mathcal{B}$.*

(ii) *The function $W_{\mathcal{B}}$ is the Wigner distribution of a generalized Gaussian:*
$W_{\mathcal{B}} = W(\widehat{T}(z_0)\psi_{\mathcal{B}})$ *with*

$$\psi_M(x) = \left(\frac{\det X}{(\pi\hbar)^n}\right)^{1/4} e^{-\frac{1}{2\hbar}Mx\cdot x}, \tag{15.12}$$

*where $M$ is given by*

$$M = X + iY = L^{-2} - iP. \tag{15.13}$$

**Proof.** (i) In view of the identity

$$W(\widehat{T}(z_0)\psi_M)(z) = W\psi_M(z - z_0)$$

it is no restriction to assume that $z_0 = 0$. Let $S$ and $S'$ in $\mathrm{Sp}(n)$ be such that

$$\mathcal{B} = S(B^{2n}(\sqrt{\hbar})) = S'(B^{2n}(\sqrt{\hbar})).$$

By homogeneity we will then have $S^{-1}S'(B^{2n}(r)) = B^{2n}(r)$ for all radii $r$ and hence $S^{-1}S'$ preserves every ball centered at the origin; it must thus be a rotation and since it is also symplectic we therefore have $S^{-1}S' = U \in U(n)$. We have $S' = SU$ hence $(S'^T)^{-1}(S')^{-1} = (S^T)^{-1}S^{-1}$ and we are done. (ii) In view of (15.10) we have

$$(S_0^T)^{-1}S_0^{-1} = \begin{pmatrix} L^{-2} + PL^2P & -PL^2 \\ -L^2P & L^2 \end{pmatrix}.$$

Writing as in (10.14)

$$G = \begin{pmatrix} X + YX^{-1}Y & YX^{-1} \\ X^{-1}Y & X^{-1} \end{pmatrix}$$

we get formula (15.13). □

**Remark 205.** The result above shows that quantum blobs are obtained from each other by simple symplectic transforms of the type $S = V_{-P}M_L$; equivalently generalized Gaussians are obtained from each other by the corresponding local metaplectic operators $\widehat{S} = \widehat{V_{-P}}\widehat{M_L}$.

### 15.3.2. Averaging over quantum blobs

Positivity questions are, as we have seen when we discussed mixed states in Chapter 14, difficult questions. We are going to see that the Gaussian average of a symbol $a \geq 0$ over an ellipsoid with symplectic capacity $\geq \frac{1}{2}h$ always leads to a positive operator; this is a geometric version of the results about the Husimi distribution that we discussed in Chapter 8.

Let $\Sigma$ be a positive definite real symmetric $2n \times 2n$ matrix. To $\Sigma$ we associate the normalized Gaussian $\rho_\Sigma$ defined by

$$\rho_\Sigma(z) = \left(\frac{1}{2\pi\hbar}\right)^n (\det\Sigma)^{-1/2} e^{-\frac{1}{2\hbar}\Sigma^{-1}z \cdot z}.$$

A straightforward calculation shows that $\rho_\Sigma$ has integral 1 (and is hence probability density); computing the Fourier transform of $\rho_\Sigma * \rho_{\Sigma'}$ one moreover easily verifies that the functions $\rho_\Sigma$ satisfy the convolution identity

$$\rho_\Sigma * \rho_{\Sigma'} = \rho_{\Sigma+\Sigma'}.$$

**Proposition 206.** *Assume that $a \geq 0$. Let $\Omega_\Sigma$ be the ellipsoid $(1/2)\Sigma^{-1}Z *$ $Z \leq 1$. If $c(\Omega_\Sigma) \geq (1/2)h$ then the operator $\widehat{A_\Sigma} = \mathrm{Op_W}(a * \rho_\Sigma)$ satisfies*

$$(\widehat{A_\Sigma}\psi|\psi)_{L^2} \geq 0$$

*for all $\psi \in \mathcal{S}(\mathbb{R}^n)$.*

**Proof.** In view of Proposition 49 in Chapter 4 we have

$$(\widehat{A_\Sigma}\psi, \psi)_{L^2} = \int_{\mathbb{R}^{2n}} (a * \rho_\Sigma)(z)W\psi(z)dz,$$

hence, taking into account the fact that $\rho_\Sigma$ is an even function,

$$(\widehat{A_\Sigma}\psi, \psi)_{L^2(\mathbb{R}^n)} = \int_{\mathbb{R}^{4n}} a(u)\rho_\Sigma(z-u)W\psi(z)dudz$$

$$= \int_{\mathbb{R}^{2n}} a(u)\left(\int_{\mathbb{R}^{2n}} \rho_\Sigma(u-z)W\psi(z)dz\right)du$$

$$= \int_{\mathbb{R}^{2n}} a(u)(\rho_\Sigma * W\psi)(u)du.$$

We have $\rho_\Sigma * W\psi \geq 0$ (see Remark 105 in Chapter 8) hence the result. $\qquad\square$

## Main Formulas in Chapter 15

| | |
|---|---|
| Quantum blob | $\mathcal{B} = S(B^{2n}(\sqrt{\hbar}))$ , $S \in \mathrm{ASp}(n)$ |
| Symplectic capacity | $c(B^{2n}(\sqrt{\hbar})) = \pi\hbar = \frac{1}{2}h$ |
| Quantum phase space | $\mathrm{QPS}(2n)$ |
| Centered quantum blobs | $\mathrm{QPS}_0(2n) \equiv \mathrm{Sp}(n)/U(n)$ |
| Dimension | $\dim \mathrm{QPS}_0(2n) = n(n+1)$ |
| General case | $\mathrm{QPS}(2n) \equiv \mathbb{R}^{n(n+3)}$ |
| Dimension | $\dim \mathrm{QPS}(2n) = n(n+3)$ |
| Reduction result | $\mathcal{B} = V_{-P}M_L(B^{2n}(\sqrt{\hbar}))$, $L = L^T$ |

# Appendix A

# Sp($n$) and Mp($n$)

For details and complete proofs we refer to [33, 39] and the references therein.

## A.1. The Symplectic Group

We will work with real $2n \times 2n$ matrices written in "block form"

$$S = \begin{pmatrix} A & B \\ C & D \end{pmatrix}, \tag{A.1}$$

where each of the entries $A$, $B$, $C$, $D$ is an $n \times n$ matrix. The transpose of $S$ is as follows:

$$S^T = \begin{pmatrix} A^T & C^T \\ B^T & D^T \end{pmatrix}. \tag{A.2}$$

The standard symplectic matrix $J$ is defined by:

$$J = \begin{pmatrix} 0 & I \\ -I & 0 \end{pmatrix},$$

where $0 = 0_{n \times n}$ and $I = I_{n \times n}$. We have $\det J = 1$ and

$$J^2 = -I, \quad J^T = -J = J^{-1}.$$

**Definition A.1.** We will say that the matrix $S$ is *symplectic* if

$$S^T J S = S J S^T = J. \tag{A.3}$$

**Remark A.2.** The conditions $SJS^T = J$ and $S^T JS = J$ are in fact equivalent as can be seen by the following simple argument: the inverse of a symplectic matrix is also symplectic (see below), hence $SJS^T = J$ is equivalent to $S^{-1}J(S^{-1})^T = J$; replacing $S$ with $S^T$ (which is also symplectic: see below) this is in turn equivalent to $(S^{-1})^T JS^{-1} = J$ that is to $S^T JS$.

The matrix $J$ is itself obviously symplectic, and so is the identity matrix. Note that it immediately follows from (A.3) that a symplectic matrix has determinant $\pm 1$, and is hence invertible. We actually have the stronger result:

$$S \ symplectic \Longrightarrow \det \ S = +1.$$

A straightforward calculation, using (A.3) and the invertibility of $S$, shows that a matrix (A.1) is symplectic if and only if any of the three sets of equivalent conditions below holds:

$$\begin{cases} A^T C, D^T B \text{ symmetric}, A^T D - C^T B = I, \\ AB^T, CD^T \text{ symmetric}, AD^T - BC^T = I, \\ DC^T, AB^T \text{ symmetric}, DA^T - CB^T = I. \end{cases} \quad (A.4)$$

It follows, in particular, that the inverse of a symplectic matrix (A.1) is given by

$$S^{-1} = \begin{pmatrix} D^T & -B^T \\ -C^T & A^T \end{pmatrix}. \quad (A.5)$$

Note that in the case $n = 1$ where $A, B, C, D$ are real numbers, the conditions (A.4) reduce to $AD - BC = 1$, i.e. $\det(S) = 1$ (the proof in the general case is much less straightforward).

The conditions (A.3) can be expressed in terms of the standard symplectic form on $\mathbb{R}^{2n}$.

**Definition A.3.** The *standard symplectic form* on phase space $\mathbb{R}^{2n}$ is the antisymmetric bilinear form defined by

$$\sigma(z, z') = Jz \cdot z'. \quad (A.6)$$

If $z = (x, p)$, $z' = (x', p')$ we thus have

$$\sigma(x, p; x', p') = p \cdot x' - p' \cdot x,$$

where the dot $\cdot$ denotes the standard scalar product of vectors in $\mathbb{R}^2$.

The number $\sigma(z, z')$ is called the *symplectic product* (or *skew-product*) of the vectors $z$ and $z'$. It is sometimes also denoted by $z \wedge z'$. The symplectic product has the following immediate interpretation in the language of differential forms: $\sigma(z, z')$ is the value on $(z, z')$ of the 2-form

$$dp \wedge dx = dp_1 \wedge dx_1 + \cdots + dp_n \wedge dx_n$$

which we will identify with $\sigma$.

Observe that when $n = 1$ the number $\sigma(z, z')$ is just minus the determinant of the vectors $z$ and $z'$: $\sigma(z, z') = -\det(z, z')$. In the general case $\sigma(z, z')$ can be expressed, using determinants, as

$$\sigma(x, p; x', p') = -\sum_{j=1}^{n} \begin{vmatrix} x_j & x'_j \\ p_j & p'_j \end{vmatrix}.$$

It is thus the algebraic sum of the areas of the projections of the vectors $z, z'$ on the conjugate coordinates axes: $pdx = p_1 dx_1 + \cdots + p_n dx_n$:

$$\sigma = d(pdx) = dp \wedge dx.$$

Since the condition $S^T J S = J$ is equivalent to $(Sz)^T J (Sz') = z^T J z'$ for all $z$, $z'$, a $2n \times 2n$ matrix $S$ is symplectic if and only if we have

$$\sigma(Sz, Sz') = \sigma(z, z') \tag{A.7}$$

for all vectors $z, z' \in \mathbb{R}^{2n}$. In terms of the differential form $\sigma = dp \wedge dx$ this can be written $S^* \sigma = \sigma$, where the upper star $*$ denotes the "pull-back" of differential forms by diffeomorphisms.

**Definition A.4.** A symplectic matrix written in block-form (A.1) is said to be free if $\det B \neq 0$; the generating function of the free symplectic matrix is the quadratic form

$$\mathcal{A}(x, x') = \frac{1}{2} B^{-1} A x^2 - B^{-1} x \cdot x' + \frac{1}{2} D B^{-1} x'^2. \tag{A.8}$$

The main interest of free symplectic matrices comes from the fact that they are entirely determined by their generating function:

$$(x, p) = S(x', p') \iff \begin{cases} p = \partial_x \mathcal{A}(x, x') \\ p' = -\partial_{x'} \mathcal{A}(x, x') \end{cases}$$

as is easily verified by a direct calculation. We will therefore write $S = S_{\mathcal{A}}$. The inverse $(S_{\mathcal{A}})^{-1}$ is also a free symplectic matrix, and we have

$$(S_{\mathcal{A}})^{-1} = S_{\mathcal{A}^*}, \quad \mathcal{A}^*(x, x') = -\mathcal{A}(x', x). \tag{A.9}$$

The unitary group $U(n, \mathbb{C})$ is identified with a subgroup $U(n)$ of $\mathrm{Sp}(n)$ by the monomorphism

$$A + iB \longmapsto \begin{pmatrix} A & -B \\ B & A \end{pmatrix}$$

and we have

$$U \in U(n) \Longleftrightarrow UJ = JU. \tag{A.10}$$

The group $U(n)$ consists of *symplectic rotations*:

$$U(n) = \mathrm{Sp}(n) \cap O(2n, \mathbb{R}). \tag{A.11}$$

The elements of $U(n)$ can be identified with the block-matrices

$$U = \begin{pmatrix} A & -B \\ B & A \end{pmatrix},$$

where $A$ and $B$ satisfy the conditions

$$AB^T = B^T A, \quad AA^T + BB^T = I. \tag{A.12}$$

## A.2. The Metaplectic Group

The symplectic group $\mathrm{Sp}(n)$ is connected, and its first homotopy group $\pi_1[\mathrm{Sp}(n)]$ is (isomorphic to) the integer group $\mathbb{Z}$. It follows from the elementary theory of covering spaces that $\mathrm{Sp}(n)$ has covering groups $\mathrm{Sp}_q(n)$ of all orders $q = 2, 3, \ldots, +\infty$, and $\mathrm{Sp}_\infty(n)$ is the universal (= simply connected) covering. The double cover $\mathrm{Sp}_2(n)$ can be faithfully represented by a group of unitary operators acting on $L^2(\mathbb{R}^n)$, the metaplectic group $\mathrm{Mp}(n)$. Let us describe the elements of $\mathrm{Mp}(n)$. We have seen above that the $\mathrm{Sp}(n)$ is generated by the free symplectic matrices

$$S = \begin{pmatrix} A & B \\ C & D \end{pmatrix}, \quad \det B \neq 0.$$

To each such matrix we associated the generating function

$$\mathcal{A}(x, x') = \frac{1}{2} DB^{-1} x^2 - B^{-1} x \cdot x' + \frac{1}{2} B^{-1} A x'^2.$$

Conversely, to every polynomial of the type

$$\mathcal{A}(x, x') = \frac{1}{2} P x^2 - L x \cdot x' + \frac{1}{2} Q x'^2$$

with $P = P^T$, $Q = Q^T$, and $\det L \neq 0$, $\tag{A.13}$

we can associate a free symplectic matrix, namely

$$S_{\mathcal{A}} = \begin{pmatrix} L^{-1}Q & L^{-1} \\ PL^{-1}Q - L^T & PL^{-1} \end{pmatrix}. \tag{A.14}$$

We now associate an operator $\widehat{S}_{\mathcal{A},m}$ to every $S_{\mathcal{A}}$ by setting, for $\psi \in \mathcal{S}(\mathbb{R}^n)$,

$$\widehat{S}_{\mathcal{A},m}\psi(x) = \left(\frac{1}{2\pi i\hbar}\right)^{n/2} \Delta(\mathcal{A}) \int_{\mathbb{R}^n} e^{\frac{i}{\hbar}\mathcal{A}(x,x')}\psi(x')dx'; \tag{A.15}$$

here $\arg i = \pi/2$ and the factor $\Delta(\mathcal{A})$ is defined by

$$\Delta(\mathcal{A}) = i^m \sqrt{|\det L|}; \tag{A.16}$$

the integer $m$ corresponds to a choice of $\arg \det L$:

$$m\pi \equiv \arg \det L \mod 2\pi. \tag{A.17}$$

Notice that we can rewrite definition (A.15) in the form

$$\widehat{S}_{\mathcal{A},m}\psi(x) = \left(\frac{1}{2\pi\hbar}\right)^{n/2} \left(e^{-i\frac{\pi}{4}}\right)^{\mu} \Delta(\mathcal{A}) \int_{\mathbb{R}^n} e^{\frac{i}{\hbar}\mathcal{A}(x,x')}\psi(x')dx', \tag{A.18}$$

where $\mu = 2m - n$.

Since generating function of $J$ being simply $\mathcal{A}(x,x') = -x \cdot x'$, it follows that

$$\widehat{J}\psi(x) = \left(\frac{1}{2\pi i\hbar}\right)^{n/2} \int_{\mathbb{R}^n} e^{-\frac{i}{\hbar}x \cdot x'}\psi(x')dx' = i^{-n/2}F\psi(x) \tag{A.19}$$

for $\psi \in \mathcal{S}(\mathbb{R}^n)$; $F$ is here the usual unitary Fourier transform defined by

$$F\psi(x) = \left(\frac{1}{2\pi\hbar}\right)^{n/2} \int_{\mathbb{R}^n} e^{-\frac{i}{\hbar}x \cdot x'}\psi(x')dx'.$$

It follows from the Fourier inversion formula

$$F^{-1}\psi(x) = \left(\frac{1}{2\pi\hbar}\right)^{n/2} \int_{\mathbb{R}^n} e^{\frac{i}{\hbar}x \cdot x'}\psi(x')dx'$$

that the inverse $\widehat{J}^{-1}$ of $\widehat{J}$ is given by the formula

$$\widehat{J}^{-1}\psi(x) = \left(\frac{i}{2\pi\hbar}\right)^{n/2} \int_{\mathbb{R}^n} e^{ix \cdot x'}\psi(x')dx' = i^{n/2}F^{-1}\psi(x).$$

Defining operators $\widehat{V}_{-P}$ and $\widehat{M}_{L,m}$ by

$$\widehat{V}_{-P}\psi(x) = e^{\frac{i}{2}Px\cdot x}\psi(x), \quad \widehat{M}_{L,m}\psi(x) = i^m\sqrt{|\det L|}\psi(Lx). \qquad (A.20)$$

we have the following useful factorization result:

$$\widehat{S}_{A,m} = \widehat{V}_{-P}\widehat{M}_{L,m}\widetilde{J}\widetilde{V}_{-Q}; \qquad (A.21)$$

it follows that the operators $\widehat{S}_{A,m}$ extend to unitary operators $L^2(\mathbb{R}^n) \longrightarrow L^2(\mathbb{R}^n)$ and the inverse of $\widehat{S}_{A,m}$ is

$$\widehat{S}_{A,m}^{-1} = \widehat{S}_{A^*,m^*} \text{ with } \mathcal{A}^*(x,x') = -\mathcal{A}(x',x), m^* = n - m. \qquad (A.22)$$

**Proposition A.5.** *Every $\widehat{S} \in \mathrm{Mp}(n)$ can be written as a product of exactly two quadratic Fourier transforms: $\widehat{S} = \widehat{S}_{A,m}\widehat{S}_{A',m'}$. (Such a factorization is, however, never unique.)*

**Corollary A.6.** *The metaplectic group $\mathrm{Mp}(2n,\mathbb{R})$ is generated by the operators $\widehat{V}_{-P}$, $\widehat{M}_{L,m}$, and $\widehat{J}$.*

### A.3.  The Inhomogeneous Metaplectic Group

We denote by $\mathrm{T}(n)$ the group of phase space translations: $T(z_0) \in \mathrm{T}(n)$ is the mapping $z \longmapsto z + z_0$. The group $\mathrm{T}(n)$ is isomorphic to the additive group $\mathbb{R}^{2n}$.

**Definition A.7.** The affine (or inhomogeneous) symplectic group is the semi-direct product

$$\mathrm{ASp}(n) = \mathrm{Sp}(n) \ltimes \mathrm{T}(n).$$

Formally, the group law of the semidirect product $\mathrm{ASp}(n)$ is given by

$$(S,z)(S',z') = (SS', z + Sz');$$

this is conveniently written in matrix form as

$$\begin{pmatrix} S & z \\ 0_{1\times 2n} & 1 \end{pmatrix} \begin{pmatrix} S' & z' \\ 0_{1\times 2n} & 1 \end{pmatrix} = \begin{pmatrix} SS' & Sz' + z \\ 0_{1\times 2n} & 1 \end{pmatrix}. \qquad (A.23)$$

One immediately checks that $\mathrm{ASp}(n)$ is identified with the set of all affine transformations $F$ of $\mathbb{R}^n \oplus \mathbb{R}^n$ such that $F$ can be factorized as a product $F = ST(z)$ for some $S \in \mathrm{Sp}(2n,\mathbb{R})$ and $z \in \mathbb{R}^{2n}$. Since translations are symplectomorphisms in their own right, it follows that $\mathrm{ASp}(n)$ is the group of all affine symplectomorphisms of $\mathbb{R}^{2n}$. We note that the transformations $T(z)$ satisfy the intertwining relations

$$ST(z) = T(Sz)S, \quad T(z)S = ST(S^{-1}z).$$

The notion of generating function also makes sense for affine symplectic mappings.

**Proposition A.8.** *Let* $F = T(z_0)S_{\mathcal{A}} \in \mathrm{ASp}(n)$.

(i) *A free generating function of* $T(z_0)S_{\mathcal{A}}$ *is the function*

$$\mathcal{A}_{z_0}(x, x') = \mathcal{A}(x - x_0, x') + p_0 \cdot x, \qquad (A.24)$$

*where* $z_0 = (x_0, p_0)$.
(ii) *Conversely, if* $\mathcal{A}$ *is the generating function of* $S_{\mathcal{A}}$ *then any polynomial*

$$\mathcal{A}_{z_0}(x, x') = \mathcal{A}(x, x') + \alpha \cdot x + \alpha' \cdot x' \qquad (A.25)$$

*with* $\alpha, \alpha' \in \mathbb{R}^n$ *is a generating function of an affine symplectic transformation* $T(z_0)S_{\mathcal{A}}$ *with* $z_0 = (x_0, p_0) = (B\alpha, D\alpha + \beta)$.

**Proof.** Let $\mathcal{A}_{z_0}$ be defined by (A.24), and set $(x', p') = S(x'', p'')$, $(x, p) = T(z_0)(x', p')$. We have

$$pdx - p'dx' = (pdx - p''dx'') + (p''dx'' - p'dx')$$
$$= (pdx - (p - p_0)d(x - x_0) + d\mathcal{A}(x'', x')$$
$$= d(p_0 \cdot x + \mathcal{A}(x - x_0, x')),$$

which shows that $\mathcal{A}_{z_0}$ is a generating function. Finally, formula (A.25) is obtained by a direct computation, expanding the quadratic form $\mathcal{A}(x - x_0, x')$ in its variables. $\square$

**Corollary A.9.** *Let* $[S_{\mathcal{A}}, z_0]$ *be a free affine symplectic transformation, and set*

$$(x, p) = [S_{\mathcal{A}}, z_0](x', p').$$

*The function* $\Phi_{z_0}$ *defined by*

$$\Phi_{z_0}(x, x') = \frac{1}{2}p \cdot x - \frac{1}{2}p' \cdot x' + \frac{1}{2}\sigma(z, z_0) \qquad (A.26)$$

*is also a free generating function for $f$; in fact*:

$$\Phi_{z_0}(x, x') = \mathcal{A}_{z_0}(x, x') + \frac{1}{2}p_0 \cdot x_0. \tag{A.27}$$

**Proof.** Setting $(x'', p'') = S(x, p)$, the generating function $\mathcal{A}$ satisfies

$$\mathcal{A}(x'', x') = \frac{1}{2}p'' \cdot x'' - \frac{1}{2}p' \cdot x'$$

in view of Euler's formula for homogeneous functions. Let $\Phi_{z_0}$ be defined by formula (A.26); in view of (A.24) we have

$$\mathcal{A}_{z_0}(x, x') - \Phi_{z_0}(x, x') = \frac{1}{2}p_0 \cdot x - \frac{1}{2}p \cdot x_0 - \frac{1}{2}p_0 \cdot x_0$$

which is (A.27); this proves the corollary since all generating functions of a symplectic transformation are equal up to an additive constant. $\qquad \square$

To each $T(z_0)S_{\mathcal{A}} \in \mathrm{ASp}(n)$ we can attach the operator $\widehat{T}(z_0)\widehat{S}_{\mathcal{A},m}$. These operators generate a group of unitary operators, the inhomogeneous metaplectic group $\mathrm{IM_p}(n)$; by definition the generating function of $T(z_0)S_{\mathcal{A}}$ is the quadratic polynomial (A.27).

# Appendix B

# The Symplectic Fourier Transform

Recall that the usual (unitary) $\hbar$-Fourier transform on the $m$-dimensional Euclidean space $\mathbb{R}^m$ is defined by

$$Ff(u) = \left(\frac{1}{2\pi\hbar}\right)^m \int_{\mathbb{R}^m} e^{-\frac{i}{\hbar}u\cdot v} f(v) dv. \tag{B.1}$$

We now assume $m = 2n$.

**Definition B.1.** The symplectic Fourier transform $F_\sigma$ on $\mathbb{R}^{2n}$ is defined, for $a \in \mathcal{S}(\mathbb{R}^{2n})$, by

$$F_\sigma a(z) = a_\sigma(z) = \left(\frac{1}{2\pi\hbar}\right)^n \int_{\mathbb{R}^{2n}} e^{-\frac{i}{\hbar}\sigma(z,z')} a(z') dz', \tag{B.2}$$

where $\sigma$ is the standard symplectic form on $\mathbb{R}^{2n}$.

The following properties of the symplectic Fourier transform easily follow from the general theory of the Fourier transform (see [39]).

- $F_\sigma$ and the Fourier transform $F$ on $\mathbb{R}^{2n}$ are related by the formula

$$F_\sigma a(z) = Fa(Jz) = F(a \circ J)(z). \tag{B.3}$$

- $F_\sigma$ is an involutive and unitary automorphism of $L^2(\mathbb{R}^{2n})$:

$$F_\sigma \circ F_\sigma = I, \quad ||F_\sigma a||_{L^2(\mathbb{R}^{2n})} = ||a||_{L^2(\mathbb{R}^{2n})}. \tag{B.4}$$

The symplectic Fourier transform $F_\sigma$ is thus its own inverse: $F_\sigma^{-1} = F_\sigma$ on $S'(\mathbb{R}^{2n})$:

$$F_\sigma a(z) = \left(\frac{1}{2\pi\hbar}\right)^n \int_{\mathbb{R}^{2n}} e^{-\frac{i}{\hbar}\sigma(z,z')} a(z')dz',$$

$$a(z) = \left(\frac{1}{2\pi\hbar}\right)^n \int_{\mathbb{R}^{2n}} e^{-\frac{i}{\hbar}\sigma(z,z')} F_\sigma a(z')dz'.$$

More generally, the symplectic Fourier transform behaves well under the action of the symplectic group:

- For $a \in S'(\mathbb{R}^{2n})$ and $S \in \mathrm{Sp}(n)$, we have

$$F_\sigma a(Sz) = F_\sigma(a \circ S)(z). \tag{B.5}$$

This immediately follows from the equalities

$$F_\sigma a(Sz) = \left(\frac{1}{2\pi\hbar}\right)^n \int_{\mathbb{R}^{2n}} e^{-\frac{i}{\hbar}\sigma(Sz,z')} a(z')dz'$$

$$= \left(\frac{1}{2\pi\hbar}\right)^n \int_{\mathbb{R}^{2n}} e^{-\frac{i}{\hbar}\sigma(z,S^{-1}z')} a(z')dz'$$

$$= \left(\frac{1}{2\pi\hbar}\right)^n \int_{\mathbb{R}^{2n}} e^{-\frac{i}{\hbar}\sigma(z,z')} a(Sz')dz',$$

where we have used the relation $\sigma(Sz, z') = \sigma(z, S^{-1}z')$ and the fact that $\det S = 1$.

- We have

$$(F_\sigma a | b)_{L^2(\mathbb{R}^{2n})} = (a | F_\sigma b)_{L^2(\mathbb{R}^{2n})}. \tag{B.6}$$

This is a straightforward consequence of the fact that $F_\sigma$ is a unitary involution:

$$(F_\sigma a | b)_{L^2(\mathbb{R}^{2n})} = (F_\sigma^2 a | F_\sigma b)_{L^2} = (a | F_\sigma b)_{L^2}.$$

Recall the usual Plancherel formula: for $a, b \in L^2(\mathbb{R}^{2n})$ we have

$$\int_{\mathbb{R}^{2n}} a(z)\overline{b(z)}dz = \int_{\mathbb{R}^{2n}} Fa(z)\overline{Fb(z)}dz \tag{B.7}$$

or, equivalently,

$$\int_{\mathbb{R}^{2n}} a(z)\overline{b(z)}dz = \int_{\mathbb{R}^{2n}} F_\sigma a(z)\overline{F_\sigma b(z)}dz, \qquad (B.8)$$

$$\int_{\mathbb{R}^{2n}} a(z)b(z)dz = \int_{\mathbb{R}^{2n}} F_\sigma a(z)F_\sigma b(-z)dz. \qquad (B.9)$$

As the usual Fourier transform, $F_\sigma$ satisfies the convolution formula

$$F_\sigma(a*b) = (2\pi\hbar)^n (F_\sigma a)(F_\sigma a) \qquad \qquad (B.10)$$

and hence

$$F_\sigma(ab) = \left(\frac{1}{2\pi\hbar}\right)^n (F_\sigma a)*(F_\sigma a). \qquad (B.11)$$

# Symplectic Diagonalization

## C.1. Williamson's Theorem

Let $M$ be a positive definite and symmetric real $2n \times 2n$ matrix. A famous theorem due to Williamson says that $M$ can be diagonalized using a symplectic matrix.

**Proposition C.1.** *There exists $S \in \mathrm{Sp}(n)$ such that*

$$S^T M S = \begin{pmatrix} \Lambda & 0 \\ 0 & \Lambda \end{pmatrix}, \tag{C.1}$$

*where $\Lambda$ is the diagonal matrix*

$$\Lambda = \begin{pmatrix} \lambda_1 & 0 & \cdots & 0 \\ 0 & \lambda_2 & \cdots & 0 \\ \vdots & \vdots & \ddots & \vdots \\ 0 & 0 & \cdots & \lambda_n \end{pmatrix}, \tag{C.2}$$

*the numbers $\lambda_j$ being the moduli of the eigenvalues of $JM$ (these are of the type $\pm i\lambda_j$ with $\lambda_j > 0$ since they are identical to those of the antisymmetric matrix $M^{1/2} J M^{1/2}$).*

The positive numbers $\lambda_1, \lambda_2, \ldots, \lambda_n$ are called the *symplectic eigenvalues* of $M$. There are several proofs of Williamson's theorem. We follow here the one we have given in [39] following Folland [28]. Let $\langle \cdot, \cdot \rangle_M$ be the complex scalar product on $\mathbb{C}^{2n}$ defined by $\langle z, z' \rangle_M = \langle Mz, z' \rangle$. Since both $\langle \cdot, \cdot \rangle_M$ and the symplectic form are non-degenerate, we can find a unique invertible matrix $K$ of order $2n$ such that

$$\langle z, Kz' \rangle_M = \sigma(z, z')$$

for all $z, z'$; that matrix satisfies $K^T M = J = -MK$. Since the $\sigma$ is antisymmetric we must have $K = -K^M$ where $K^M = -M^{-1} K^T M$ is the transpose of $K$ with respect to $\langle \cdot, \cdot \rangle_M$; it follows that the eigenvalues of $K = -M^{-1} J$ are of the type $\pm i\lambda_j$, $\lambda_j > 0$, and so are those of $JM^{-1}$. The corresponding complex eigenvectors occurring in conjugate pairs $e'_j \pm if'_j$ we thus obtain a $\langle \cdot, \cdot \rangle_M$-orthonormal basis $\{e'_i, f'_j\}_{1 \leq i,j \leq n}$ of $\mathbb{R}^n \oplus \mathbb{R}^n$ such that $Ke'_i = \lambda_i f'_i$ and $Kf'_j = -\lambda_j e'_j$. Notice that it follows from these relations that we have $K^2 e'_i = -\lambda_i^2 e'_i$ and $K^2 f'_j = -\lambda_j^2 f'_j$ and that the vectors of the basis $\{e'_i, f'_j\}_{1 \leq i,j \leq n}$ satisfy the relations

$$\sigma(e'_i, e'_j) = \langle e'_i, Ke'_j \rangle_M = \lambda_j \langle e'_i, f'_j \rangle_M = 0,$$

$$\sigma(f'_i, f'_j) = \langle f'_i, Kf'_j \rangle_M = -\lambda_j \langle f'_i, e'_j \rangle_M = 0,$$

$$\sigma(f'_i, e'_j) = \langle f'_i, Ke'_j \rangle_M = \lambda_i \langle f'_i, f'_j \rangle_M = -\lambda_i \delta_{ij}.$$

Setting $e_i = \lambda_i^{-1/2} e'_i$ and $f_j = \lambda_j^{-1/2} f'_j$, the basis $\{e_i, f_j\}_{1 \leq i,j \leq n}$ is symplectic. Let $S$ be the element of $\mathrm{Sp}(2n, \mathbb{R})$ mapping the canonical symplectic basis to $\{e_i, f_j\}_{1 \leq i,j \leq n}$. The $\langle \cdot, \cdot \rangle_M$-orthogonality of $\{e_i, f_j\}_{1 \leq i,j \leq n}$ implies (C.1) with $\Lambda$ given by (C.2).

The symplectic matrix $S$ diagonalizing $M$ in Williamson's theorem is unique up to a symplectic rotation: if $S'$ is another Williamson diagonalizing symplectic matrix then $S^{-1}S' \in U(n)$. To prove this, assume that $S' \in \mathrm{Sp}(n)$ is such that

$$S'^T M S' = S^T M S = D = \begin{pmatrix} \Lambda & 0 \\ 0 & \Lambda \end{pmatrix}$$

and set $U = S^{-1}S'$; we have $U^T DU = D$. Let us show that $UJ = JU$; it will follow that $U \in U(n)$. Setting $R = D^{1/2} U D^{-1/2}$ we have

$$R^T R = D^{-1/2}(U^T DU)D^{-1/2} = D^{-1/2} DD^{-1/2} = I$$

hence $R \in O(2n)$. Since $J$ commutes with each power of $D$ we have, since $JU = (U^T)^{-1}J$ (because $U$ is symplectic),

$$JR = D^{1/2} JU D^{-1/2} = D^{1/2}(U^T)^{-1} J D^{-1/2}$$

$$= D^{1/2}(U^T)^{-1} D^{-1/2} J = (R^T)^{-1} J$$

hence $R \in \mathrm{Sp}(2n, \mathbb{R}) \cap O(2n)$; in view of (A.11) we thus have $R \in U(n)$ hence $JR = RJ$. Now $U = D^{-1/2}RD^{1/2}$ and therefore

$$
\begin{aligned}
JU &= JD^{-1/2}RD^{1/2} = D^{-1/2}JRD^{1/2} \\
&= D^{-1/2}RJD^{1/2} = D^{-1/2}RD^{1/2}J \\
&= UJ
\end{aligned}
$$

which was to be proven.

## C.2. The Block-Diagonal Case

When the matrix $M$ is block-diagonal, one has the following more precise result.

**Proposition C.2.** *Let $A$ and $B$ be two positive-definite symmetric $n \times n$ matrices. There exists $L \in GL(n, \mathbb{R})$ such that*

$$
L^T A L = L^{-1}B(L^T)^{-1} = \Lambda, \tag{C.3}
$$

*where $\Lambda$ is the diagonal matrix*

$$
\Lambda = \begin{pmatrix} \sqrt{\lambda_1} & 0 & \cdots & 0 \\ 0 & \sqrt{\lambda_2} & \cdots & 0 \\ \vdots & \vdots & \ddots & \vdots \\ 0 & 0 & \cdots & \sqrt{\lambda_n} \end{pmatrix}, \tag{C.4}
$$

*the positive numbers $\lambda_1, \ldots, \lambda_n$ being the eigenvalues of $AB$.*

**Proof.** We are following [39]. That the eigenvalues of $AB$ are positive follows from the fact that since $A$ and $B$ are positive definite matrices then the eigenvalues of $AB$ are those of the positive definite matrix $A^{1/2}BA^{1/2}$. We claim that there exists $R \in GL(n, \mathbb{R})$ such that

$$
R^T A R = I \quad \text{and} \quad R^{-1}B(R^T)^{-1} = D, \tag{C.5}
$$

where

$$
D = \begin{pmatrix} \lambda_1 & 0 & \cdots & 0 \\ 0 & \lambda_2 & \cdots & 0 \\ \vdots & \vdots & \ddots & \vdots \\ 0 & 0 & \cdots & \lambda_n \end{pmatrix}.
$$

In fact, first choose $P \in \mathrm{GL}(n, \mathbb{R})$ such that $P^T A P = I$ and set $B_1^{-1} = P^T B^{-1} P$. Since $B_1^{-1}$ is symmetric, there exists $H \in O(n, \mathbb{R})$ (the orthogonal group) such that $B_1^{-1} = H^T D^{-1} H$ where $D^{-1}$ is diagonal. Set now $R = PH^T$; we have $R^T A R = I$ and also

$$R^{-1} B (R^T)^{-1} = H P^{-1} B (P^T)^{-1} H^T = H B_1 H^T = D;$$

the equalities (C.5) follow. Let $\Lambda$ be the diagonal matrix (C.4). Since

$$R^T A B (R^T)^{-1} = R^T A R (R^{-1} B (R^T)^{-1}) = D,$$

the diagonal elements of $D$ are indeed the eigenvalues of $AB$ hence $D = \Lambda^2$. Setting $L = R\Lambda^{1/2}$ we have

$$L^T A L = \Lambda^{1/2} R^T A R \Lambda^{1/2} = \Lambda,$$

$$L^{-1} B (L^{-1})^T = \Lambda^{-1/2} R^{-1} B (R^T)^{-1} \Lambda^{-1/2} = \Lambda$$

hence our claim.                                                                 $\square$

### C.3.  The Symplectic Case

When the matrix $M$ is itself symplectic, we have the following result.

**Proposition C.3.** *Let $S \in \mathrm{Sp}(n)$ be symmetric and positive definite. There exists $U \in U(n)$ such that*

$$S = U^T \Delta U, \quad \Delta = \begin{pmatrix} \Lambda & 0 \\ 0 & \Lambda^{-1} \end{pmatrix}, \tag{C.6}$$

*where*

$$\Lambda = \begin{pmatrix} \lambda_1 & 0 & \cdots & 0 \\ 0 & \lambda_2 & \cdots & 0 \\ \vdots & \vdots & \ddots & \vdots \\ 0 & 0 & \cdots & \lambda_n \end{pmatrix}$$

*and $0 < \lambda_1 \leq \lambda_2 \leq \cdots \leq \lambda_n \leq 1$ are the $n$ first eigenvalues of $S$ (counted with their multiplicities).*

**Proof.** Since $S$ is symmetric and positive definite, its eigenvalues occur in pairs $(\lambda, 1/\lambda)$ with $\lambda > 0$ so that if $\lambda_1 \leq \cdots \leq \lambda_n$ are $n$ eigenvalues then $1/\lambda_1, \ldots, 1/\lambda_n$ are the other $n$ eigenvalues. Let now $U$ be an orthogonal matrix such that $S = U^T \Delta U$ where $\Delta$ is as in (C.6). Let us prove that $U \in U(n)$. Let $u_1, \ldots, u_n$ be orthonormal eigenvectors of $U$ corresponding to the eigenvalues $\lambda_1, \ldots, \lambda_n$. Since $SJ = JS^{-1}$ (because $S$ is both symplectic and symmetric) we have

$$SJu_k = JS^{-1}u_k = \frac{1}{\lambda_j}Ju_k$$

for $1 \leq k \leq n$, hence $\pm Ju_1, \ldots, \pm Ju_n$ are the orthonormal eigenvectors of $U$ corresponding to the remaining $n$ eigenvalues $1/\lambda_1, \ldots, 1/\lambda_n$. Write now the $2n \times n$ matrix $[u_1, \ldots, u_n]$ as

$$[u_1, \ldots, u_n] = \begin{pmatrix} A \\ B \end{pmatrix},$$

where $A$ and $B$ are of order $n \times n$; we have

$$(-Ju_1, \ldots, -Ju_n) = -J \begin{pmatrix} A \\ B \end{pmatrix} = \begin{pmatrix} -B \\ A \end{pmatrix}$$

hence $U$ is of the type

$$U = (u_1, \ldots, u_n; -Ju_1, \ldots, -Ju_n) = \begin{pmatrix} A & -B \\ B & A \end{pmatrix}.$$

Since $U^T U = I$ the blocks $A$ and $B$ are such that

$$AB^T = B^T A, \quad AA^T + BB^T = I \tag{C.7}$$

hence $U$ is also symplectic, that is $U \in U(n)$. $\square$

A related question is whether one can find other symplectic diagonalizations of $M$ where $\Lambda$ is replaced by some diagonal other matrix $\Lambda'$ whose diagonal entries are numbers $\lambda_j$ different from the $\lambda_j^\sigma$. The answer is negative: we must have $\Lambda = \Lambda'$ up to a reordering of the entries (see [33, §8.3.1, Theorem 8.11(ii)] for a proof).

## C.4. The Symplectic Spectrum

The symplectic spectrum of a positive definite symmetric real matrix $M$ is defined as follows.

**Definition C.4.** Let $\lambda_1, \lambda_2, \ldots, \lambda_n$ be the symplectic eigenvalues of $M$. The positive numbers $\lambda_j$ are the "symplectic eigenvalues" of the positive definite matrix $M$. Ordering these eigenvalues decreasingly: $\lambda_1 \geq \lambda_2 \geq \cdots \geq \lambda_n$, the symplectic spectrum of $M$ is the ordered set

$$\text{Spec}_\sigma(M) = (\lambda_1, \ldots, \lambda_n).$$

Let us list and prove the main properties of the symplectic spectrum:

**Proposition C.5.** *Let $M$ and $M'$ be a real positive-definite matrix of order $2n$. The symplectic spectrum $\mathrm{Spec}_\sigma(M)$ has the following properties:*

(i) $\mathrm{Spec}_\sigma(M)$ *is a symplectic invariant:* $\mathrm{Spec}_\sigma(S^T M S) = \mathrm{Spec}_\sigma(M)$ *for every $S \in \mathrm{Sp}(n)$.*

(ii) *If* $\quad \mathrm{Spec}_\sigma(M) \quad = \quad (\lambda_1, \ldots, \lambda_n), \quad$ *then* $\mathrm{Spec}_\sigma(M^{-1}) = ((\lambda_n)^{-1}, \ldots, (\lambda_1)^{-1})$.

(iii) *If $M \le M'$, then $\mathrm{Spec}_\sigma(M) \le \mathrm{Spec}_\sigma(M')$, i.e. we have $\lambda_j \le \lambda'_j$ for all $j = 1, \ldots, n$.*

**Proof.** (i) This follows from the fact that $J(S^T M S) = S^{-1}(JM)S$ (because $JS^T = S^{-1}J$ since $S$ is symplectic) which implies that $JM$ and $J(S^T M S)$ have the same eigenvalues.

(ii) The eigenvalues of $JM$ are the same as those of $M^{1/2}JM^{1/2}$; the eigenvalues of $JM^{-1}$ are those of $M^{-1/2}JM^{-1/2}$. Now

$$M^{-1/2}JM^{-1/2} = -(M^{1/2}JM^{1/2})^{-1}$$

hence the eigenvalues of $JM$ and $JM^{-1}$ are obtained from each other by the transformation $t \longmapsto -1/t$. The result follows since the symplectic spectra are obtained by taking the moduli of these eigenvalues.

(iii) Notice that when $A$ or $B$ is invertible $AB$ and $BA$ have same eigenvalues. We will write $M \le M'$ when $M' - M$ is non-negative; equivalently, $M \le M'$ is equivalent to $z^T M z \le z^T M' z$ for every $z \in \mathbb{R}^{2n}$. We now observe that the statement (ii) is equivalent to

$$M \le M' \Longrightarrow (JM')^2 \le (JM)^2$$

since the eigenvalues of $JM$ and $JM'$ occur in pairs $\pm i\lambda$, $\pm i\lambda'$ with $\lambda$ and $\lambda'$ real. Replacing $z$ by successively $(JM^{1/2})z$ and $(JM'^{1/2})z$ in $z^T M z \le z^T M' z$ we thus have, taking into account the fact that $J^T = -J$, that is, since $J^T = -J$,

$$M^{1/2}JM'JM^{1/2} \le M^{1/2}JMJM^{1/2}. \tag{C.8}$$

$$M'^{1/2}JM'JM'^{1/2} \le M'^{1/2}JMJM'^{1/2}. \tag{C.9}$$

Noting that we have $M^{1/2}JM'JM^{1/2} \simeq MJM'J$ and $M'^{1/2}JMJM'^{1/2} \simeq M'JMJ \simeq MJM'J$ (where $\simeq$ is the relation "has the same eigenvalues as") and we can rewrite the relations (C.8) and (C.9) as

$$MJM' \leq JM^{1/2}JM'JM^{1/2} , M'^{1/2}JM'JM'^{1/2} \leq MJM'J$$

and hence, by transitivity,

$$M'^{1/2}JM'JM'^{1/2} \leq M^{1/2}JMJM^{1/2}. \qquad (C.10)$$

Since we have $M^{1/2}JMJM^{1/2} \simeq (MJ)^2$ and $M'^{1/2}JM'JM'^{1/2} \simeq (M'J)^2$ the relation (C.10) is equivalent to $(M'J)^2 \leq (MJ)^2$, which was to be proven. $\qquad \square$

Appendix D

# Symplectic Capacities

Symplectic capacities were introduced by Ekeland and Hofer [21]. The existence of symplectic capacities is guaranteed by Gromov's symplectic non-squeezing theorem [55]. For a detailed account see [46].

## D.1. Gromov's Non-squeezing Theorem

A symplectomorphism of $\mathbb{R}^{2n}$ (equipped with its standard symplectic form $\sigma$) is a diffeomorphism $f$ of $\mathbb{R}^{2n}$ such that the Jacobian matrix $Df(z)$ at every $z \in \mathbb{R}^{2n}$ is symplectic: $Df(z) \in \mathrm{Sp}(n)$. Let us denote by $B^{2n}(R)$ the ball in $\mathbb{R}^{2n}$ with radius $R$, and by $Z_j^{2n}(R)$ the cylinder in $\mathbb{R}^{2n}$ defined by $x_j^2 + p_j^2 \leq R^2$. Gromov proved the following very deep theorem from symplectic topology.

**Theorem D.1 (Gromov).** *If there exists a symplectomorphism $f$ of $\mathbb{R}^{2n}$ sending the phase space ball $B^{2n}(r)$ in some cylinder $Z_j^{2n}(R)$, then we must have $r \leq R$.*

Gromov's theorem (which is often referred to as "the Principle of the Symplectic camel") allows two define two symplectic capacities: the Gromov width, and the cylindrical capacity.

**Definition D.2.** (i) The Gromov width $c_G(\Omega)$ of a subset $\Omega$ of $\mathbb{R}^{2n}$ is given by

$$c_G(\Omega) = \sup_f \{\pi R^2 : f(B^{2n}(R)) \subset \Omega\}. \tag{D.1a}$$

(ii) The cylindrical capacity $c_Z(\Omega)$ is given by

$$c_Z(\Omega) = \inf_f \{\pi R^2 : f(\Omega) \subset Z_j^{2n}(R)\}. \tag{D.1b}$$

In both cases $f$ ranges over the group of all symplectomorphisms of $\mathbb{R}^{2n}$.

Notice that Gromov's non-squeezing theorem implies that

$$c_G(B^{2n}(R)) = c_G(Z_j^{2n}(R)) = \pi R^2,$$

$$c_Z(B^{2n}(R)) = c_Z(Z_j^{2n}(R)) = \pi R^2.$$

## D.2. Symplectic Capacities

The following definition extends the notion of Gromov's width.

**Definition D.3.** A symplectic capacity on $(\mathbb{R}^{2n}, \sigma)$ is a function assigning to every subset $\Omega$ of $\mathbb{R}^{2n}$ a number $c(\Omega) \geq 0$, or $+\infty$, and having the following properties:

(**SC1**) It must be invariant under symplectomorphisms:

$$c(f(\Omega)) = c(\Omega) \quad \text{if } f \text{ is a symplectomorphism.} \tag{D.2}$$

(**SC2**) It must be monotone with respect to set inclusion:

$$c(\Omega) \leq c(\Omega') \quad \text{if } \Omega \subset \Omega'. \tag{D.3}$$

(**SC3**) It must behave like an area under dilations:

$$c(\lambda\Omega) = \lambda^2 c(\Omega) \quad \text{for any scalar } \lambda \tag{D.4}$$

($\lambda\Omega$ is the set of all points $\lambda z$ such that $z \in \Omega$).
(**SC4**) It satisfies the normalization conditions

$$c(B^{2n}(R)) = \pi R^2 = c(Z_j^{2n}(R)). \tag{D.5}$$

One shows that the Gromov width and the cylindrical capacity are, respectively, the smallest and the largest symplectic capacity:

$$c_G(\Omega) \leq c(\Omega) \leq c_Z(\Omega)$$

for all symplectic capacities $c$. There are infinitely many symplectic capacities: for every real $\lambda$ in the closed interval $[0, 1]$ the formula

$$c_\lambda = \lambda c_G + (1 - \lambda)c_Z \qquad (D.6)$$

obviously defines a symplectic capacity, and we have $c_\lambda \neq c_{\lambda'}$ if $\lambda \neq \lambda'$. More generally, we can always interpolate two arbitrary symplectic capacities to obtain new capacities.

We also have the weaker notion of linear symplectic capacity.

**Definition D.4.** A "linear symplectic capacity" assigns to every subset $\Omega$ of $\mathbb{R}^{2n}$ a number $c^{\mathrm{lin}}(\Omega) \geq 0$ or $+\infty$, such that the properties (SC2)–(SC4) above hold, while property (SC1) is replaced with the following:

(**SC1Lin**) A linear symplectic capacity $c^{\mathrm{lin}}$ is invariant under phase-space translations and under the action of $\mathrm{Sp}(n)$.

Property (SC1Lin) can be restated by saying that $c^{\mathrm{lin}}$ invariant under the action of the semidirect product $\mathrm{ASp}(n) = \mathrm{Sp}(n) \ltimes \mathrm{T}(n)$ of the symplectic group and of the translation group ($\mathrm{ASp}(n)$ is the affine symplectic group).

## D.3. Properties

Obviously symplectic capacities are unbounded (even if the symplectic capacity of an unbounded set can be bounded, cf. property (SC4)). We have for instance

$$c(\mathbb{R}^{2n}) = +\infty$$

as immediately follows from the formula $c(B^{2n}(R)) = \pi R^2$. However, if $\Omega$ is bounded then all its symplectic capacities are finite: there exists $R$ (perhaps very large) such that a ball $B^{2n}(R)$ contains $\Omega$, and one then concludes using the monotonicity property (SC2) which implies that we have:

$$c(\Omega) \leq c(B^{2n}(R)) = \pi R^2.$$

In fact, more generally, it follows from the monotonicity property and from (SC4) that, if

$$B^{2n}(R) \subset \Omega \subset Z_j^{2n}(R)$$

then $c(\Omega) = \pi R^2$; this illustrates the fact that sets very different in shape and volume can have the same symplectic capacity.

There exist infinitely many symplectic capacities.

The notion of symplectic capacity is not directly related to that of volume. For instance, the function $c_{\mathrm{Vol}}$ defined by

$$c_{\mathrm{Vol}}(\Omega) = [\mathrm{Vol}(\Omega)]^{1/n}$$

obviously satisfies the properties (SC1)–(SC4) above, except of course the identity

$$c_{\mathrm{Vol}}(Z_j^{2n}(R)) = \pi R^2$$

(as soon as $n > 1$) which is precisely the most characteristic and interesting property of a symplectic capacity. Also, $c(\Omega) = \mathrm{Area}(\Omega)$ when $\Omega$ is a connected and simply connected surface in the plane.

### D.4. The Symplectic Capacity of an Ellipsoid

The following result is very important; it shows that all symplectic capacities (linear, or not) agree on phase space ellipsoids.

**Proposition D.5.** *Let $M$ be a $2n \times 2n$ positive-definite matrix $M$ and $\lambda_1, \ldots, \lambda_n$ its symplectic eigenvalues. Consider the ellipsoid:*

$$\Omega_M : Mz \cdot z \leq 1. \tag{D.7}$$

(i) *For every symplectic capacity $c$ on $(\mathbb{R}^{2n}, \sigma)$ we have*

$$c(\Omega_M) = \frac{\pi}{\lambda_{\max}}, \tag{D.8}$$

*where $\lambda_{\max} = \lambda_1$ is the largest symplectic eigenvalue of $M$.*

(ii) *In particular, if $A$ and $B$ are real symmetric $n \times n$ matrices, then the symplectic capacity of the ellipsoid*

$$\Omega_{(A,B)} : Ax^2 + Bp^2 \leq 1 \tag{D.9}$$

*is given by*

$$c(\Omega_{(A,B)}) = \frac{\pi}{\sqrt{\lambda_{\max}}}, \tag{D.10}$$

*where $\lambda_{\max}$ is the largest eigenvalue of $AB$.*

**Proof.** The statement (ii) follows from (i) in view of Proposition C.2 in Appendix C, choosing

$$M = \begin{pmatrix} A & 0 \\ 0 & B \end{pmatrix}.$$

To prove (i) we first note that in view of Williamson's theorem (Appendix C) there exists a symplectic matrix $S$ such that

$$S^T M S = \begin{pmatrix} \Lambda & 0 \\ 0 & \Lambda \end{pmatrix}, \tag{D.11}$$

where $\Lambda$ is the diagonal matrix with diagonal elements, the symplectic eigenvalues $\lambda_j$ of $M$. Since symplectic capacities are invariant by canonical transformations, it follows that $c(\Omega_M) = c(S(\Omega_M))$ hence it suffices to prove formula (D.8) when $\Omega_M$ is replaced by $S(\Omega_M)$. We may thus assume that

$$\Omega_M : \sum_{j=1}^{n} \frac{1}{R_j^2}(x_j^2 + p_j^2) \leq 1, \tag{D.12}$$

where we have set $\lambda_j = 1/R_j^2$. Suppose now that there exists a canonical transformation $f$ sending a ball $B^{2n}(R)$ inside $\Omega_M$. Then $f(B^{2n}(R))$ is also contained in each cylinder $Z_j^{2n}(R) : x_j^2 + p_j^2 \leq R^2$ and hence $R \leq R_{\min} = \sqrt{1/\lambda_{\max}}$ in view of Gromov's non-squeezing theorem (Appendix D). It follows that $c_{\min}(\Omega_M) \leq \pi R_{\min}^2 = \pi/\lambda_{\max}$; since on the other hand $B^{2n}(R_{\max})$ is anyway contained in $\Omega_M$ we must have equality: $c_{\min}(\Omega_M) = \pi/\lambda_{\max}$. A similar argument shows that we also have $c_{\max}(\Omega_M) = \pi/\lambda_{\max}$; formula (D.8) follows since $c_{\min}$ and $c_{\max}$ are the smallest and largest symplectic capacities. $\square$

It is often necessary to consider the dual ellipsoid of a given ellipsoid. Let $Q$ be a positive definite quadratic form on $\mathbb{R}^{2n}$ and consider the real function $f_\zeta$ defined for $\zeta \in \mathbb{R}^{2n}$ by $f_\zeta(z) = z \cdot \zeta - Q(z)$. That function has a unique critical value, which we denote by $Q^*(\zeta)$. The function $Q^*$ thus defined is the *Legendre transform* of $Q$; it is also a positive definite quadratic form.[2] In fact, a straightforward calculation shows that if $Q(z) = Mz^2$ where $M$ is a positive-definite symmetric matrix then $Q^*(\zeta) = \frac{1}{4}M^{-1}\zeta^2$. In particular if $Q$ is the Hessian matrix of $Q$ then the Hessian matrix of $Q^*$ is the inverse $Q^{-1}$. We thus have the useful formula:

$$Q(z) = \frac{1}{2}Qz \cdot z \Longleftrightarrow Q^*(\zeta) = \frac{1}{2}Q^{-1}\zeta \cdot \zeta. \tag{D.13}$$

**Definition D.6.** Let $\Omega_M$ be the phase-space ellipsoid defined by $Q(z) \leq 1$ where $Q$ is a positive-definite quadratic form on $\mathbb{R}^{2n}$. The dual ellipsoid

---

[2] $Q^*$ is *stricto sensu* defined on the dual space of $\mathbb{R}^{2n}$ but we will gladly ignore this technicality here.

$\Omega_M^*$ of $\Omega_M$ is defined by $Q^*(\zeta) \leq 1$ where $Q^*$ is the Legendre transform of $Q$.

Note that it immediately follows from (D.13) that we have $(\Omega_M^*)^* = \Omega_M$. The following result gives the symplectic capacity of the dual ellipsoid.

**Proposition D.7.** *For $M > 0$ let $\Omega_M : Mz^2 \leq 1$ and $\Omega_M^* : \frac{1}{4}M^{-1}\zeta^2 \leq 1$ be dual phase-space ellipsoids. Let $\mathrm{Spec}_\sigma(M) = (\lambda_1, \ldots, \lambda_n)$ be the symplectic spectrum of $M$. We have*

$$c(\Omega_M^*) = \frac{\pi}{4}\lambda_{\min}, \tag{D.14}$$

*where $\lambda_{\min}$ is the smallest symplectic eigenvalue of $M$.*

**Proof.** The dual ellipsoid $\Omega_M^*$ is determined by $2M^{-1}\zeta^2 \leq 1$. Using Proposition C.5 (ii) in Appendix C we have

$$\mathrm{Spec}_\sigma((4M)^{-1}) = ((4\lambda_n)^{-1}, \ldots, (4\lambda_1)^{-1})$$

hence, applying formula (D.8) for the symplectic capacity of an ellipsoid $c(\Omega_M^*) = \pi/4(\lambda_n)^{-1}$ which is precisely formula (D.14) since $\lambda_n$ is the smallest symplectic eigenvalue of $M$. $\square$

# Bibliography

[1] F. Agorram, A. Benkhadra, A. El Hamyani and A. Ghanmi, Complex Hermite functions as Fourier–Wigner transform, *Integral Transforms Special Funct.*, **27**(2), 94–100 (2015).

[2] Y. Aharonov, P. G. Bergmann and J. Lebowitz, Time symmetry in the quantum process of measurement, *Phys. Rev. B*, **134**, B1410–B1416 (1964).

[3] Y. Aharonov, E. Cohen and A. C. Elitzur, Can a future choice affect a past measurement's outcome? *Ann. Phys.*, **355**, 258–268 (2015).

[4] F. Bayen, M. Flato, C. Fronsdal, A. Lichnerowicz and D. Sternheimer, Deformation theory and quantization. I. Deformation of symplectic structures, *Ann. Phys.*, **110**, 111–151 (1978).

[5] F. Bayen, M. Flato, C. Fronsdal, A. Lichnerowicz and D. Sternheimer, Deformation theory and quantization. II. Physical applications, *Ann. Phys.*, **11**, 6–110 (1978).

[6] S. Bochner, *Lectures on Fourier Integrals*, Princeton University Press, Princeton, NJ, 1959.

[7] P. Boggiatto and Elena Cordero, Anti-Wick quantization with symbols in $L^p$ spaces, *Proc. Amer. Math. Soc.*, **130**(9), 2679–2685 (2002).

[8] P. Boggiatto, G. De Donno and A. Oliaro, Time-frequency representations of Wigner type and pseudo-differential operators, *Trans. Amer. Math. Soc.*, **362**(9), 4955–4981 (2010).

[9] P. Boggiatto, B. K. Cuong, G. De Donno and A. Oliaro, Weighted integrals of Wigner representations, *J. Pseudo-Differential Operators Appl.*, **1**(4), 401–415 (2010).

[10] P. Boggiatto, G. Donno and A. Oliaro, Hudson's theorem for $\tau$-Wigner transforms, *Bull. London Math. Soc.*, **45**(6), 1131–1147 (2013).

[11] F. Bopp, La mécanique quantique est-elle une mécanique statistique partic-
     ulière? *Ann. Inst. H. Poincaré*, **15**, 81–112 (1956).

[12] M. Born and P. Jordan, Zur Quantenmechanik, *Zeits. Physik*, **34**, 858–888
     (1925).

[13] T. Bröcker and R. F. Werner, Mixed states with positive Wigner functions,
     *J. Math. Phys.*, **36**(1), 62–75 (1995).

[14] N. G. De Bruijn, A theory of generalized functions, with applications to
     Wigner distribution and Weyl correspondence, *Nieuw Arch. Wiskd.*, **21**(3),
     205–280 (1973).

[15] E. Cordero, M. de Gosson and F. Nicola, On the Invertibility of Born-Jordan
     Quantization, *J. Math. Pures Appl.*, **05**(4), 537–557 (2016).

[16] J. P. Dahl, On the group of translations and inversions of phase space and
     the Wigner functions, *Physica Scripta*, **25**(4), 499–503 (1982).

[17] G. Dennis, M. A. de Gosson and B. J. Hiley, Bohm's quantum potential as
     an internal energy, *Phys. Lett. A*, **379**(18), 1224–1227 (2015).

[18] N. Dias, M. de Gosson, J. Prata, Maximal covariance group of Wigner
     transforms and pseudo-differential operators, *Proc. Amer. Math. Soc.*,
     **142**(9), 3183–3192 (2014).

[19] N. Dias and J. Prata, The Narcowich–Wigner spectrum of a pure state, *Rep.
     Math. Phys.*, **63**(1), 43–54 (2009).

[20] J. Du and M. W. Wong, A trace formula for Weyl transforms, *Approx.
     Theory. Appl. (N.S.)*, **16**(1), 41–45 (2000).

[21] I. Ekeland and H. Hofer, Symplectic topology and Hamiltonian dynamics,
     I and II. *Math. Z.*, **200**, 355–378 and **203**, 553–567 (1990).

[22] H. G. Feichtinger, On a new Segal algebra, *Monatsh. Math.*, **92**(4), 269–289
     (1981).

[23] H. G. Feichtinger, Banach spaces of distributions of Wiener's type and
     interpolation, in *Functional Analysis and Approximation, Oberwohlfach*,
     International Series of Numerical Mathematics, Vol. 60, Birkhäuser, Basel,
     pp. 153–165 (1981).

[24] H. G. Feichtinger, Modulation spaces: Looking back and ahead sample,
     *Theory Signal Image Process*, **5**(2), 109–140 (2006).

[25] R. Flack and B. J. Hiley, Weak measurement and its experimental realisa-
     tion, *J. Phys. Conf. Ser.*, **504**(1), 012016. IOP Publishing, 2014.

[26] P. Flandrin, Maximum signal energy concentration in a time-frequency
     domain, *Acoustics, Speech, Signal Process.*, **4**, 2176–2179 (1988).

[27] G. B. Folland, *Harmonic Analysis in Phase Space*, Annals of Mathematics
     Studies, Princeton University Press, Princeton, NJ, 1989.

[28] J. M. Garcia-Bondia and J. C. Várilly. Nonnegative mixed states in Weyl–
     Wigner–Moyal theory, *Phys. Lett. A*, **128**(1–2), 20–24 (1988).

[29] M. de Gosson, *The Principles of Newtonian and Quantum Mechanics*,
     Imperial College Press, London, 1st edition (2001), 2nd edition (2016).

[30] M. de Gosson, Phase space quantization and the uncertainty principle. *Phys.
     Lett. A*, **317**(5), 365–369 (2003).

[31] M. de Gosson, The optimal pure Gaussian state canonically associated to a
     Gaussian quantum state, *Phys. Lett. A*, **330**(3), 161–167 (2004).

[32] M. de Gosson, *Symplectic Geometry and Quantum Mechanics*, Operator Theory: Advances and Applications (subseries: "Advances in Partial Differential Equations"), Vol. 166, Birkäuser, Basel, (2006).

[33] M. de Gosson, Extended Weyl calculus and application to the phase-space Schrödinger equation, *J. Phys. A*, **38**(19), L325–L329 (2005).

[34] M. de Gosson, Symplectically covariant Schrödinger equation in phase space, *J. Phys. A*, **38**(42), 9263–9287 (2005).

[35] M. de Gosson, Weyl Calculus in phase space and the Torres–Vega and Frederick equation, in *Quantum Theory: Reconsideration of Foundations*, AIP Conference Proceedings, Vol. 810, American Institute of Physics., Melville, NY, (2006) Vol. 3, pp. 300–304.

[36] M. de Gosson, Spectral properties of a class of generalized landau operators, *Comm. Partial Differential Operators*, **33**(11), 2096–2104 (2008).

[37] M. de Gosson, The symplectic camel and the uncertainty principle: the tip of an iceberg? *Found. Phys.*, **39**(2), 194–214 (2009).

[38] M. de Gosson, *Symplectic Methods in Harmonic Analysis and in Mathematical Physics*, Birkhäuser, 2011.

[39] M. de Gosson, Quantum blobs, *Found. Phys.*, **43**(4), 440–457 (2013).

[40] M. de Gosson, *Introduction to Born–Jordan Quantization*, Fundamental Theories of Physics, Springer (2016).

[41] M. de Gosson, From Weyl to Born–Jordan quantization: The Schrödinger representation revisited, *Phys. Rep.*, **623**, 1–58 (2016).

[42] C. de Gosson and M. de Gosson, The phase space formulation of time-symmetric quantum mechanics, *Quanta*, **4**(1), 27–34 (2015).

[43] M. de Gosson and F. Luef, Quantum states and Hardy's formulation of the uncertainty principle: A symplectic approach, *Lett. Math. Phys.*, **80**, 69–82 (2007).

[44] M. de Gosson and F. Luef, Remarks on the fact that the uncertainty principle does not determine the quantum state, *Phys. Lett. A*, **364**(6), 453–457 (2007).

[45] M. de Gosson and F. Luef, Symplectic capacities and the geometry of uncertainty: The irruption of symplectic topology in classical and quantum mechanics, *Phys. Rep.*, **484**, 131–179 (2009).

[46] M. de Gosson and F. Luef, Born–Jordan Pseudodifferential calculus, Bopp operators and deformation quantization, *Integral Equations Operator Theory*, **84**(4) 463–485 (2016).

[47] I. S. Gradshteyn and I. M. Ryzhik, *Table of Integrals, Series, and Products*, Academic press, New York, 1980.

[48] B. J. Hiley, On the relationship between the Wigner–Moyal and Bohm approaches to quantum mechanics: A step to a more general theory? *Found. Phys.*, **40**(4), 356–367 (2010).

[49] B. J. Hiley, Weak values: Approach through the Clifford and Moyal algebras, *J. Phys. Conf. Ser.*, **361**(1), (2012).

[50] B. J. Hiley, On the relationship between the Wigner–Moyal approach and the quantum operator algebra of von Neumann, *J. Comput. Electron.*, **14**(4), 869–878 (2015).

[51] K. Gröchenig, *Foundations of Time-Frequency Analysis*, Birkhäuser, Boston, 2000.

[52] K. Gröchenig and C. Heil, Modulation spaces and pseudodifferential operators, *Integral Equations Operator Theory*, **34**, 439–457 (1999).

[53] H. J. Groenewold, On the principles of elementary quantum mechanics, *Physica*, **12**(7), 405–460 (1946).

[54] M. Gromov, Pseudoholomorphic curves in symplectic manifolds, *Invent. Math.*, **82**, 307–347 (1985).

[55] A. Grossmann, Parity operators and quantization of δ-functions,*Commun. Math. Phys.*, **48**, 191–193 (1976).

[56] G. H. Hardy, A theorem concerning Fourier transforms, *J. London Math. Soc.*, **8**, 227–231 (1933).

[57] M. O. S. Hillery, R. F. O'Connell, M. Scully and E. P. Wigner, Distribution functions in physics: fundamentals, *Phys. Rep.*, (Review Section of Physics Letters), **106**(3), 121–167 (1984).

[58] R. L. Hudson, When is the Wigner quasi-probability density non-negative? *Rep. Math. Phys.*, **6**, 249–252 (1974).

[59] A. J. E. M. Janssen, A Note on Hudson's theorem about functions with nonnegative wigner distributions, *SIAM J. Math. Anal.*, **15**(1), 170–176 (1984).

[60] A. J. E. M. Janssen, Positivity and spread of bilinear time-frequency distributions, in *The Wigner Distribution*, Elsevier, Amsterdam, (1997), pp. 1–58.

[61] D. Kastler, The $C^*$-algebras of a free boson field, *Commun. Math. Phys.*, **1**, 14–48 (1965).

[62] Y. Katznelson, *An Introduction to Harmonic Analysis*, Dover, New York, 1976.

[63] E. H. Lieb and Y. Ostrover, Localization of multidimensional Wigner distributions, *J. Math. Phys.*, **51**(10), 102101 (2010).

[64] G. Loupias and S. Miracle-Sole, $C^*$-algèbres des systèmes canoniques, I, *Commun. Math. Phys.*, **2**, 31–48 (1966).

[65] G. Loupias and S. Miracle-Sole, $C^*$-algèbres des systèmes canoniques, II, *Ann. Inst. Henri Poincaré*, **6**(1), 39–58 (1967).

[66] J. M. Maillard, On the twisted convolution product and the Weyl transformation of tempered distributions, *J. Geom. Phys.*,**3**(2), 231–261 (1986).

[67] J. Moyal, Quantum mechanics as a statistical theory. *Math. Proc. Cambridge Philos. Soc.*, **45**(99), 99–124 (1949).

[68] F. J. Narcowich, Conditions for the convolution of two Wigner distributions to be itself a Wigner distribution, *J. Math. Phys.*, **29**(9), 2036–2041 (1988).

[69] F. J. Narcowich, Distributions of ℏ-positive type and applications, *J. Math. Phys.*, **30**(11), 2565–2573 (1989).

[70] F. J. Narcowich, Geometry and uncertainty, *J. Math. Phys.*, **31**(2), 354–364 (1990).

[71] F. J. Narcowich and R. F. O'Connell, Necessary and sufficient conditions for a phase-space function to be a Wigner distribution, *Phys. Rev. A*, **34**(1), 1–6 (1986).

[72] F. J. Narcowich and R. F. O'Connell, A unified approach to quantum dynamical maps and Gaussian Wigner distributions, *Phys. Lett. A*, **133**(4) 167–170 (1988).

[73] M. Reed and B. Simon, *Methods of Modern Mathematical Physics*, Academic Press, New York, 1972.

[74] A. Royer, Wigner functions as the expectation value of a parity operator, *Phys. Rev. A*, **15**, 449–450 (1977).

[75] M. A. Shubin, *Pseudodifferential Operators and Spectral Theory*, Springer-Verlag, 1987 [original Russian edition in Nauka, Moskva, 1978].

[76] F. Soto and P. Claverie, When is the Wigner function of multidimensional systems nonnegative? *J. Math. Phys.*, **24**(1), 97–100 (1983).

[77] E. M. Stein, *Harmonic Analysis : Real-Variable Methods, Orthogonality, and Oscillatory Integrals*, Vol. 43, Princeton University Press, 2016.

[78] J. Toft, Hudson's theorem and rank one operators in Weyl calculus, in *Pseudo-differential Operators and related topics*, Birkhäuser, Basel, (2006), pp. 153–159.

[79] H. Weyl, Quantenmechanik und Gruppentheorie, *Z. Physik.*, **46**, 1–46 (1927).

[80] E. Wigner, On the quantum correction for thermodynamic equilibrium, *Phys. Rev.*, **40**, 799–755 (1932).

[81] M. W. Wong, *Weyl Transforms*, Springer, 1998.

# Index

**A**

algebra property, 85
antisymplectic, 60
auto-ambiguity function, 31

**B**

Banach Gelfand triple, 87
Bargmann transform, 73
Bochner's theorem, 169
Bopp operators, 139
Bopp shifts, 136

**C**

canonical commutation relations, 40
characteristic function, 170
chirp, 54
coboundary, 133
Cohen class, 89
covariance, 163
covariance matrix, 147, 165, 187
cross-ambiguity function, 31
cross-Wigner transform, 17

**D**

density operator, 155, 159
displacement operator, 7

distributional kernel, 43
dual of the Feichtinger algebra, 86

**E**

expectation value, 145
exponentials, 14

**F**

Feichtinger algebra, 77
fiducial coherent state, 104
Fourier–Wigner transform, 33, 69
free symplectic matrix, 54, 199

**G**

Gabor transform, 19, 73, 77
Gaussians, 103
Gelfand triple, 87
generalized Gaussian, 103
generalized Moyal identity, 94
generators of the symplectic group, 54
Grossmann–Royer operator, 9, 17, 42

**H**

$\hbar$-positivity, 179

ℏ-positive type, 171
Hardy's uncertainty principle, 119
Heisenberg inequalities, 148
Heisenberg–Weyl operator, 7, 31, 172
Hermite functions, 110
Hermite polynomials, 109
Hermite's differential equation, 109
Hilbert–Schmidt operator, 46, 156
Hudson's theorem, 109
Husimi distribution (generalized), 90,
    97

**I**

inhomogeneous metaplectic group,
    202
inversion of the Wigner transform, 70

**K**

KLM conditions, 171

**L**

Laguerre function, 110
Laguerre polynomial, 111–112

**M**

marginal conditions, 92
marginal properties of the Wigner
    transform, 26
maximal covariance, 59
metaplectic representation, 53
mixed quantum state, 155
Moyal identity, 67, 94
Moyal product, 131
multidimensional Hardy uncertainty
    principle, 121

**N**

Narcowich–Wigner spectrum, 169,
    179

**O**

orthogonal projector, 160

**P**

partial isometry, 73
Plancherel formula, 204
polarization identity, 22
purity of a quantum state, 161

**Q**

quantization, 14
quantum blob, 183, 187
quantum covariance matrix, 148, 175
quasi-distribution, 144

**R**

radar ambiguity function, 32
reconstruction formula, 70
reconstruction of the wavefunction,
    152
Robertson–Schrödinger inequalities,
    148, 164
Rodrigue's formula, 109

**S**

sesquilinear, 22
short-time Fourier transform (*see also*
    STFT), 19
spectrogram, 20
squeezed coherent state, 103–104
standard Gaussian, 108
standard symplectic form, 4, 196
standard symplectic matrix, 4
sub-Gaussian estimate, 123
symbol, 39
symplectic covariance, 55
symplectic eigenvalue, 177, 207
symplectic form, 196
symplectic Fourier transform, 134,
    171, 203
symplectic matrix, 4, 195

**T**

trace-class operator, 156
translation of Wigner transforms,
    27

twisted convolution, 134, 173
twisted symbol, 41, 173

**U**
uncertainty principle, 147, 162

**V**
variance, 163

**W**
$\eta$-Weyl operator, 181
$\eta$-Wigner transform, 180
wavepacket transform, 72, 83

weak value, 145, 150
Weyl correspondence, 39
Weyl operator, 39–40, 144
Weyl's characteristic function, 14
Wigner transform, 18
Wigner transform of a Gaussian, 105, 107
Wigner transform of Hermite functions, 109, 112
Williamson diagonalization, 207
windowed Fourier transform, 19
Woodward ambiguity function, 32

Printed in the United States
By Bookmasters